# Energy and Life

# Energy and Life

**John M. Wrigglesworth**

King's College, University of London, UK

Taylor & Francis
Publishers since 1798

**UK**  Taylor & Francis Ltd., 1 Gunpowder Square, London EC4A 3DE
**USA**  Taylor & Francis Inc., 1900 Frost Road, Suite 101, Bristol PA 19007

**British Library Cataloguing in Publication Data**

A catalogue record for this book is available from the British Library.

ISBN 0-7484-0433-3

**Library of Congress Cataloging Publication Data are available**

Cover design by Jim Wilkie

Typeset in Melior 10/12½pt by Keyword Typesetting Services, Surrey

Printed in Great Britain by T. J. International, Padstow, Cornwall

# Contents

# General Preface to the Series

The curriculum for higher education now presents most degree programmes as a collection of discrete packages or modules. The modules stand alone but, as a set, comprise a general programme of study. Usually around half of the modules taken by the undergraduate are compulsory and count as a core curriculum for the final degree. The arrangement has the advantage of flexibility. The range of options over and above the core curriculum allows the student to choose the best programme for his or her future.

Usually, the subject of the core curriculum, for example biochemistry, has a general textbook that covers the material at length. Smaller specialist volumes deal in depth with particular topics, for example photosynthesis or muscle contraction. The optional subjects in a modular system, however, are too many for the student to buy the general textbook for each and the small in-depth titles generally do not cover sufficient material. The new series *Modules in Life Sciences* provides a selection of texts which can be used at the undergraduate level for subjects optional to the main programme of study. Each volume aims to cover the material at a depth suitable to a particular year of undergraduate study with an amount appropriate to a module, usually around one quarter of the undergraduate year. The life sciences was chosen as the general subject area since it is here, more than most, that individual topics proliferate. For example, a student of biochemistry may take optional modules in physiology, microbiology, medical pathology and even mathematics.

Suggestions for new modules and comments on the present volumes will always be welcomed and should be addressed to the series editor.

*John Wrigglesworth, Series Editor*
King's College, London

# Preface

This book covers the subject of biological energetics at a breadth suitable to be taken as an individual module of study at the intermediate level of an undergraduate degree. The text takes as its theme the main mechanisms for coupling the free energy of catabolic processes to the reactions necessary for life. The approach is quantitative and detailed molecular structures are not given unless essential for numerical analysis. I make no apologies for this, although I know that numbers are often unpopular with students. I hope that this approach shows some of the beauty of the subject. It is possible to do fairly simple calculations on quite interesting problems. For example, with reasonable assumptions, the time taken for photosynthetic organisms to generate the present-day oxygen atmosphere can be estimated (within a few million years!). The food intake for reasonable body-weight balance can be calculated. It is possible to quantitate the energy needed to warm up the thoracic muscles of a bee before flight. Thermodynamics can be used to analyse living systems, but many of the parameters are unknown and the final quantitative answer may not seem sensible. Sometimes a simple numerical mistake is responsible, but once the calculations have been checked the assumptions behind the calculation need to be examined. For example, see the energy calculations on proton translocation by complex I of the respiratory chain (Figure 7.8).

The text should be suitable for non-specialists in other life-science disciplines who opt to study this topic as an optional module. Of course, depending on the background of the student, some areas will present more difficulty than others. References are therefore mainly from secondary or review literature.

I thank Harold Baum for suggesting the idea for the series and Peter Nicholls for many stimulating discussions on the subject of this volume. I am also grateful to my family for much-appreciated support and encouragement.

*John Wrigglesworth*
King's College, London

# 1 Introduction to Bioenergetics

## 1.1 Life and energy

Energy flow is essential for life and bioenergetics describes how living systems capture, transform and use energy. Almost immediately we meet a problem which turns most students away from the subject. The concept of energy is not an easy one. Definitions are very abstract, 'the capacity to do work', 'the energy of an object by virtue of its position', the 'rest-mass energy' of an object. In fact we really have no knowledge of what energy *is*. Another awkward fact is that energy also seems to exist in many different forms. We can speak of potential energy, kinetic energy, heat energy, electrical energy, chemical energy, radiant energy, nuclear energy, and even 'information' energy. Certain observational facts or laws, the laws of thermodynamics, allow us to do various calculations about energy and energy transformations but these do not lead us any closer to the abstract *thing* that is called energy. Nevertheless, the continuous flow of energy through organisms is required for life.

A second requirement for life, which is probably easier to imagine, is some method of storing information and passing the knowledge from one generation to the next. We know how this works quite well. The information is stored in the linear sequence of bases in deoxyribonucleic acid (DNA) in the form of the genetic code. Replication of DNA occurs to transmit the information from one generation to the next. The production of ribonucleic acid (RNA) (transcription) and protein (translation) allows this information to be used for the essential functions of life. Nevertheless, although an information system is necessary for life it is not sufficient on its own. For example, we do not think of viruses as living systems although they have a very efficient information storage system. They have to parasitise living cells in order to turn the viral information into a useful form. The additional factor present in living cells and missing in the virus is a mechanism for controlling a flow of energy through the system. It is interesting to note that the very bases used to mediate between the stored information in DNA and the

Two Conditions
Necessary for Life

- Mechanism(s) for the control of energy flow
- Systems for information storage and transmission

production machinery for protein synthesis are the ones present in the nucleoside triphosphates used to mediate between different forms of energy in the cell (see Box 1.1).

We shall see in later sections how energy and information are closely linked concepts. To do this we have to define

**Box 1.1   The nucleosides—an evolutionary link between information and energy flow?**

*Figure 1.1*
**Nucleoside involvement in information and energy transduction by living systems**

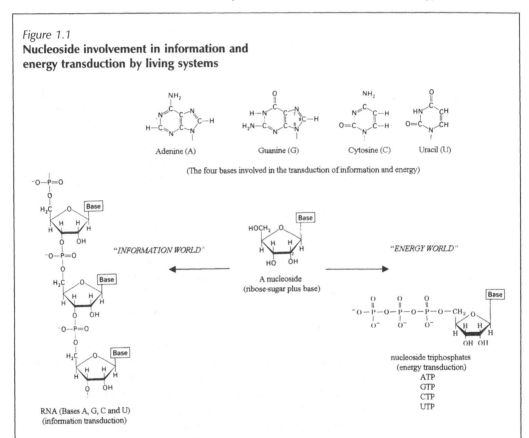

There is a fascinating story to be discovered somewhere in the early evolution of life concerning the use of nucleosides for both information transmission and energy transduction. The four nucleosides, phosphorylated to the corresponding nucleotides, are linked together in RNA to form a linear polymer with specific base sequences. The genetic code defines the translation of this information into protein. Triplet sequences in messenger RNA direct the protein-synthesising machinery to produce the specific sequence of amino acids in the primary structure of a protein.

The same four nucleosides are also used in metabolism to link energy-generating reactions with energy-utilising reactions (Figure 1.1). Each nucleoside, phosphorylated to the triphosphate form, appears to have evolved to deal with a particular area of metabolism. Thus ATP is mainly involved in reactions concerned with electrical and mechanical work (ion pumps and contractile proteins), whereas UTP is involved in carbohydrate synthesis, GTP in protein synthesis and CTP in lipid synthesis.

The evolution of nucleoside function seems to have its roots in the RNA world. The triphosphates used in metabolism contain ribose not deoxyribose, which is found in DNA, and the base uracil is used rather than thiamine, the base substituting for uracil in DNA.

more closely how energy and information are measured and what rules govern the flow or redistribution of energy in a system. It will then be possible to describe some of the molecular mechanisms by which living systems capture energy from the surroundings and redistribute it in different ways. The systems we can study in this way range from individual biochemical reactions to single cells and whole organisms. Even the growth and activity of populations can be studied in relation to the availability of energy resources.

## 1.2 Quantitation and the laws of thermodynamics

Fortunately, even though the energy of a system is very difficult to define or measure, it is relatively easy to measure *differences* in energy. For example, we can take a mole of glucose (180 g) and burn it in a bomb calorimeter in an atmosphere of oxygen. Most of the energy is given out as heat, which will raise the temperature of the surrounding material. (A small energy change also occurs because there is a change in the way in which the atoms are organised in the molecules but this need not concern us just yet.) The temperature rise is easily measured with a thermometer and, with a knowledge of the heat capacity of the surrounding material, it is a relatively simple task to calculate the number of joules or calories given out when 1 mole of glucose is oxidised to 6 moles of carbon dioxide and 6 moles of water. The number we get measures a change in energy from the initial state to the final state. It is not necessary to know exactly what energy *is* for this to be a useful measurement. This is because the first law of thermodynamics states that energy is conserved. It may be converted from one form to another but is neither created nor destroyed. Thus, any difference in energy caused by changing the bond energies and rearranging the molecular structures of glucose and oxygen has to appear as another form of energy, in this case as heat given out to the surroundings. The total energy of the system plus surroundings is the same after the reaction as before.

One consequence of the first law is that *the measured amount of energy change does not depend on how the reaction took place, only on the initial and final states*. The same number would be found if the glucose were oxidised via a large number of intermediate reactions, as long as the only products formed were carbon dioxide and water. Because of this, we will see later how calorimetry can be used to study the energy balance of whole animals.

---

The unit of energy is the **joule**, but the older unit, the **calorie**, is still used in some areas of bioenergetics. I will use the joule but give the calorie equivalent wherever appropriate (1 calorie is equivalent to 4.18 joules).

---

THE FIRST LAW

Energy is conserved. It can be changed from one form to another but it can neither be created nor destroyed.

## THE SECOND LAW

Natural processes are accompanied by an increase in the entropy of the universe.

Whereas the first law of thermodynamics is concerned with the conservation of energy, the second law (actually discovered before the first) is concerned with the distribution of energy in space. This is an important concept, probably one of the most profound that we know for living systems. Changes in this distribution with time define a direction to time, something lacking in the other laws of thermodynamics and indeed in other laws of physics. The second law defines the direction of natural change. Applying this to living systems is incredibly useful to understanding how they function. It can tell us what direction a metabolic pathway can take; how conditions could be changed to reverse a sequence of reactions; whether we become overweight if we overeat; and even why living systems evolve.

There are several statements of the second law of thermodynamics, all equivalent. The most useful for our purposes states that natural processes are accompanied by an increase in the entropy of the universe. However, just like the problem with energy, this is not really helpful unless we know what entropy is. One useful approach is to think of entropy as a measure of probability. The more probable a state is, the higher its entropy. This is explicitly stated in Boltzmann's famous equation

$$S = k \ln(W) \tag{1.1}$$

## BOLTZMANN'S CONSTANT $k$

One of the universal constants of nature at the microscopic level. It has a value of $1.381 \times 10^{-23}\,\mathrm{J\,K^{-1}}$. Multiplication by Avogadro's number $N$ $(6.022 \times 10^{23}\,\mathrm{mol^{-1}})$ gives the gas constant $R$ $(8.314\,\mathrm{J\,mol^{-1}\,deg^{-1}})$ a universal constant of the macroscopic world. $k \cdot N = R$

where $W$ is the number of ways of arranging a given distribution of microstates (see Figure 1.2). The proportionality constant $k$ is a universal constant, known as Boltzmann's constant. (Note that because probabilities are multiplied together ($W_{\mathrm{final}} = W_1 \times W_2$), the logarithmic relationship in Equation (1.1) means that entropies are additive ($S_{\mathrm{final}} = S_1 + S_2$).) The second law can then be restated as signifying that *natural processes are accompanied by an overall change to a more probable state*. This can be illustrated by reference to Figure 1.2.

It should be mentioned here that it is best not to use terms such as 'order' and 'disorder'. These soon lead to a subjective view of entropy, since one person's order is another person's disorder. A pack of playing cards in one particular sequence may be considered very ordered for one type of card game but very disordered for another. A player may recognise the presence or absence of a pattern according to whether it produces a winning hand or not. Defining entropy in terms of disorder soon leads to the apparent paradox that the subjective knowledge of a system determines its entropy. A particular system would then have a different entropy according to who was looking at it.

## 1.3 Entropy, heat and work

One of the subtle relationships between entropy and heat can also be seen from Figure 1.2. Remember that entropy is not energy but a measure of the distribution of energy. Heat is a form of energy related to molecular motion. The higher the kinetic energy of individual molecules in a mixture, the hotter that mixture is. If a hot system is placed in contact with a cold system (shown for an extreme case of energy

**Box 1.2  Probability and the second law of thermodynamics**

*Figure 1.2*
**Distribution of microstates in a system**
Initially all microstates are occupied in the subsystem (system on left). Removal of the barrier allows the occupancy of microstates to randomise (system on right)

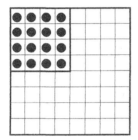 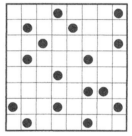

It is common sense that a highly improbable state such as the system on the left in Figure 1.2 will spontaneously change to a more probable state such as shown on the right when the subsystem barrier is removed. Initially, the dark particles fill all 16 squares of the subsystem. With only one possible arrangement, $W$, the entropy of this localised system is zero $(S = k \ln W)$. In the right-hand system the particles have been allowed to distribute at random over the available space. With many more ways of arranging their distribution, the entropy of the whole system has increased. For each particle, the number of locations it can move to is increased by a factor of 4. For 16 particles the number of possible arrangements will be $4^{16}$. The increase in entropy will therefore be $\ln(4^{16})$ or $16 \ln 4$. Any movement of a particle into the localised subsystem will reduce the number of possible arrangements in that subsystem and hence lower its entropy. Fluctuations caused by an uneven number of particles entering and leaving the subsystem will become smaller and smaller as the number of particles and the size of the system increases. The whole system will reach an equilibrium where there is a maximum probability of distribution of particles over space and time.

We can also think of the diagrams as representing distributions of energy states. For example, we could think of the initial state on the left as having a subsystem at high temperature. By redistribution, or energy flow, the whole system reaches a more probable arrangement. Eventually we get to thermal equilibrium where there is a maximum probability of energy distribution. We can then say that the temperature of the whole system is uniform. It is important to note, that, in reaching a uniform temperature, the total energy has not changed. This would be a violation of the first law of thermodynamics. All we have done is disperse the energy from a coherent to a more incoherent state. *The driving force for the change is not a lowering of energy but a dispersal of the energy into a more probable distribution.*

distribution in Figure 1.2), the kinetic energy of the molecules will be shared, by multiple collisions, and eventually a new equilibrium distribution will be formed. When there is a uniform distribution of the kinetic energy of the molecules between the systems, they are said to be in thermal equilibrium. In the initial state, the temperatures of the two systems are very different. In the final state, the temperatures of the two systems are the same. Energy has been transferred from the hot system to the cold. Thermal equilibrium is a state of maximum probability for the kinetic energy distribution of the molecules. The equilibrium state corresponds to a state of maximum entropy.

From what we have just seen, it should be apparent that entropy depends on temperature. At the high temperature, where the motion of the molecules is restricted, the entropy is low. (In fact it is zero in the artificial example given in Box 1.2.) On the other hand, at low temperatures where the distribution of heat is more random, the entropy is higher. The change in entropy of a system at temperature $T$ when a small quantity of heat ($\Delta q$) is slowly added or taken away can be expressed as

**Entropy** is a measure of the **distribution** of energy.

$$\Delta S = \frac{\Delta q}{T} \tag{1.2}$$

We make the sign of $\Delta q$ positive if heat is added so that the entropy increases. Conversely, removing heat causes the entropy of the system to decrease.

Another statement of the second law is that no process is possible in which the only result is the net transfer of energy from a cooler to a hotter body. We can use Equation (1.2) to verify this statement. Consider two systems at different temperatures $T_{hot}$ and $T_{cold}$. Let a small amount of heat be taken from the hot system and given to the cold. Equation (1.2) tells us that the entropy of the hot system will change by an amount $-\Delta q/T_{hot}$

$$\Delta S_{hot} = -\frac{\Delta q}{T_{hot}}$$

whereas the entropy of the cold system changes by $+\Delta q/T_{cold}$

$$\Delta S_{cold} = \frac{\Delta q}{T_{cold}}$$

Overall the entropy change of the universe is

$$\Delta S_{hot} + \Delta S_{cold}$$

This is always positive since $T_{hot}$ is greater than $T_{cold}$ which makes the increase in entropy $\Delta S_{cold}$ always greater than the decrease in entropy $\Delta S_{hot}$. The reverse will never happen spontaneously, since this would require the overall entropy change to be negative, contrary to the second law.

It is always possible to extract some of the energy being transferred from the hot system to the cold system in order to do work (see Box 1.3). This is the basis of how heat engines work. However, in order to keep to the requirements of the second law, we must give some of the energy in the form of heat to the cold system such that the overall change in entropy is still positive. From the previous arguments, the minimum amount of heat that has to be given to the cold system (sink) can be seen to be

$$\Delta q \, \frac{T_{cold}}{T_{hot}}$$

Thus, to maximise the amount of work energy extracted from the system, it is necessary to maximise the temperature difference between the two systems either by making $T_{cold}$ as low as possible or by making $T_{hot}$ as high as possible. The efficiency of heat engines can be expressed in terms of the amount of work extracted from the heat transferred:

*Box 1.3* **The efficiency of heat engines**

*Figure 1.3*
**Heat engines**
The engine represented by the system on the left tranduces a greater proportion of heat energy into work than the system on the right.

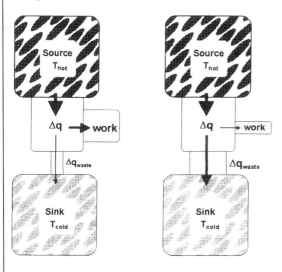

A heat engine works by transferring heat from a hot source to a cold sink. Some of the heat transferred can be used to do work and the rest enters the cold sink as randomised energy, $\Delta q_{waste}$. Different engines have different efficiencies. The engine on the left in Figure 1.3 is able to transform more of the heat into work than the one on the right. This is determined by the skill of the engineer. However, no matter how well the system is designed, the maximum amount of work that it is possible to get out of any engine is limited by the second law, and can be expressed as

$$\text{work} = \Delta q \left( 1 - \frac{T_{cold}}{T_{hot}} \right)$$

*Box 1.4*  **Work**

*Figure 1.4*
Picking up an apple by
1 metre expends
approximately 1 joule.

**lift 1 m**

**expend 1 J**

**100 g apple**

Definitions of work can be rather complicated. It can be said to be a measure of 'the displacement of extensive quantities across a difference of conjugated potentials'! The unit of work, the joule, seems not much simpler to define. It is the amount of work done when a force of 1 newton displaces the point at which it is applied by 1 metre in the direction of the force. However, Brown (1979) has placed the units of work in a more useful context. If we remember that the gravitational force pulling downwards on mass at the surface of the earth is 9.81 N, it then follows that picking up a mass of 1 kg by 1 m will expend 9.81 J. One joule would therefore be equivalent to picking up an apple from the ground (around 0.1 kg lifted by 1 m). Alternatively, a 70 kg human will use 700 J climbing up 1 m of stairway. Climbing the Eiffel tower (320 m) would require 224 kJ. This may seem a large expenditure of energy, but from Table 2.1 we can calculate that this could be met by eating around two spoonfuls of sugar.

$$\text{work} = (\text{heat transferred}) \left( 1 - \frac{T_{\text{cold}}}{T_{\text{hot}}} \right) \quad\quad (1.3)$$

Note the important conclusion that *when the temperatures of the two systems are the same, it is not possible to use heat to generate work.*

### 1.3.1  *Irreversibility*

We have seen that any spontaneous change takes place in the direction towards a state of increased probability (illustrated in Figure 1.2). It is possible to reverse the final distribution of microstates shown in Figure 1.2, but only by imposing external work on the system. For example if we think of the microstates as representing the distribution of the kinetic energy of molecules, then we could use a piston to compress the volume and reverse the reaction. We would, however, be putting energy into the system by doing the compression work. Without this input the initial reaction would be irreversible. In the example of the heat engines used above, any heat transferred to the cold system is effectively wasted in terms of its availability to do work. The energy is distributed randomly among the molecules of the cold system and, without any other input, remains in that state. The more efficient the heat engine, the less is the heat transferred to the cold system (see Box 1.3). Using Equation (1.2) we can express the waste heat in terms of an irreversible entropy change $\Delta S_{\text{irrev}}$:

---

\ ←
→ ↑ \ ↓
← ↓ → ←
↑ → / ↓
\ ↑ ← /

**Heat**, chaotic motion

→ → → →
→ → → →
→ → → →

**Work**, coherent motion
(can be mechanical, as
in the directed move-
ment of mass, or elec-
trical by the directed
movement of charge, or
chemical by the orien-
tation of electron or-
bitals in a chemical
bond)

---

$$\Delta q_{\text{waste}} = T \, \Delta S_{\text{irrev}} \tag{1.4}$$

We know from the second law that it is not possible to eliminate $\Delta q_{\text{waste}}$ unless $T_{\text{cold}}/T_{\text{hot}}$ is zero. It turns out that the same waste energy is produced in all types of reactions which occur spontaneously, not only in heat engines. The amount of waste energy, measured by $T \, \Delta S_{\text{irrev}}$ differs according to the efficiency of the reaction, but it is not possible to eliminate it altogether.

An everyday illustration of the relationship between entropy, heat and work is the stretching of a rubber band (Box 1.5). Work is done on the system to stretch the rubber. This organises the molecules into linear arrays, an improbable distribution. The drop in entropy is accompanied by release of heat $\Delta q_{\text{waste}}$, which can be felt if the stretched rubber is immediately put to your forehead. The reverse happens if the rubber is allowed to contract. Work is now done by the band—we could attach it to a pulley and make it lift a weight. The entropy of the rubber increases as the molecules resume their more random arrangements and the temperature drops. Being an open system, waste heat flows in from the surroundings. Again this can be felt if the band is put to your forehead immediately after it is allowed to contract. The use of entropy change to do work has been adapted by the flea (see Figure 1.6), this time by compressing rubber-like material. It is not possible using chemical reactions (in a safe manner!) to generate energy at a fast enough rate for the flea to jump away from its enemies. It therefore stores energy in the form of an ordered array of rubber-like molecules in its leg. A suitable trigger releases the energy and away jumps the flea. Of course the flea has to do work to re-order the molecules again but that can be done later by coupled chemical reactions.

## 1.4 Living systems

### 1.4.1 Living systems are isothermal

There are no significant temperature differences within cells. From what we have just learnt above, it would be of little use in metabolism to release heat from a chemical reaction and hope to convert that heat into work such as muscle contraction or ion gradient formation. The only use of the reaction would be to keep warm (which in itself is an important function of metabolism). For example, the hydrolysis of ATP to ADP releases a considerable amount of energy. Unless the reaction is somehow coupled to other reactions, the energy

*Box 1.5*  **Entropic work**

*Figure 1.5*
**The structure of rubber**
Stretching the rubber extends the chains and
lowers the entropy. Contraction raises the
entropy of the system which can be made to
do work.

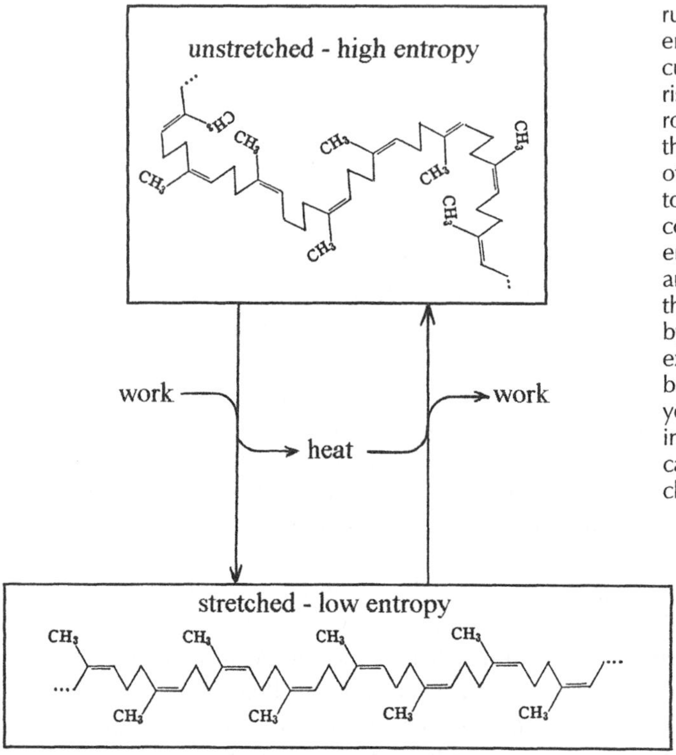

unstretched - high entropy

work

heat

work

stretched - low entropy

Natural rubber is made up
of long chainlike molecules.
When stretched, these polymers are organised into linear chains with a limited
number of possible arrangements (Figure 1.5). Work is
necessary to elongate the
rubber and the drop in
entropy of the rubber molecules is accompanied by a
rise in entropy of the surroundings (heat output). On
the other hand, contraction
of the rubber can be made
to do work. The force of
contraction is almost
entirely entropic in origin
and the rise in entropy of
the system is accompanied
by heat input. Try an
experiment with a rubber
band, putting it quickly to
your forehead after stretching or contraction (take
care!) to measure the heat
changes.

*Figure 1.6*
**The jumping flea**

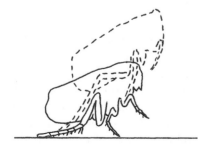

Most jumping insects have a catapult mechanism to increase the power (rate of work) of
their hind legs. The force comes from the elas-
tic energy stored in molecules of compressed
resilin. The release of the 'entropic energy' can
take place much faster than energy generated
by metabolism, although chemical reactions
are required to put the resilin under compression in the first place. (Figure adapted from
McN. Alexander, 1979, *The Invertebrates*,
Cambridge University Press. Based on draw-
ings from high-speed cine film images by H. C.
Bennet-Clark and E.C.A. Lucey, 1967, *J. Exp.
Biol.* **47**, 59-76.)

will be released as heat and effectively wasted. In order for
living systems to do work (biosynthesis of macromolecules,
formation of ion gradients across membranes, and so on)
they have to change energy from one form to another *without
using heat as an intermediate*. This is relatively straightforward if mechanisms are available, usually chemical reactions, to link the energy-generating processes to the energy-
utilising processes. A large part of this book is concerned
with such coupling mechanisms.

With heat engines, Equation (1.3) was useful to measure
the amount of energy available to do work and Equation (1.4)
gave some expression of the irreversibility of the reaction.
For living systems, where there are no temperature
differences, it is necessary to derive another function to
make it easier to relate the direction of change of a reaction
with the amount of energy available to do work.

Consider the chemical reaction

$$A \rightarrow B$$

The second law tells us that the reaction will go in the direction indicated if the total entropy change of the universe
increases during the course of the reaction. We can indicate
this as

$$\Delta S_{universe} > 0$$

It is quite possible to split the entropy term into two, the
entropy change in the reaction system, and the entropy
change in the rest of the universe. For a spontaneous reaction, we then have

$$\Delta S_{system} + \Delta S_{surroundings} > 0 \qquad (1.5)$$

Now the entropy given out to the surroundings can be
expressed in terms of the heat given out such that

$$\Delta S_{surroundings} = -\frac{\Delta q}{T}$$

In a heat engine this heat could be used to do work such as
driving a piston, but with living systems the temperature
difference between the system and the surroundings is
usually quite small, a matter of 10 or 20°C. Any work that
might be done, such as a pressure or volume change, will be
only a small fraction of the energy given out as heat. If we
ignore this, then we can then use the term *enthalpy*, given
the symbol $H$. The enthalpy of a system is a measure of the

heat change during a reaction at constant pressure. *Liberation* of heat is expressed with a minus sign. For living systems, therefore, we can approximate $-\Delta q$ as being equivalent to $-\Delta H$. Rewriting Equation (1.5) gives

$$\Delta S_{\text{system}} - \frac{\Delta H}{T} > 0$$

This is usually rearranged by multiplying both sides by $-T$ (and leaving off the subscript) to give

$$\Delta H - T\,\Delta S < 0 \tag{1.6}$$

This is a very powerful expression. It states that for a reaction to proceed in a certain direction then $\Delta H - T\,\Delta S$ has to fall during the reaction. It is the energy equivalent of the second law. The term $\Delta H - T\,\Delta S$ is called the *Gibbs free energy* or *Gibbs function* after Josiah Willard Gibbs, the first person to formulate the expression. With this function we can quantitate energy changes in chemical reactions and, more importantly, determine which direction the reaction will take. If the change in Gibbs free energy ($\Delta G$) for $A \rightarrow B$ is negative, then we know that the reaction will take place spontaneously. Conversely, if $\Delta G$ is positive then the reaction will proceed in the reverse direction. It is not a measure of how *fast* it will go, only the direction. The system may well need a catalyst to speed up the reaction.

### 1.4.2 *Living systems are non-equilibrium systems*

Even though individual reactions within a series of coupled reactions may be reversible, the binding of oxygen to haemoglobin for example, the overall change in the life of an organism is far from equilibrium. The overall direction has to be accompanied by a rise in entropy of the universe. Under isothermal conditions this means that the net change in Gibbs free energy must be negative. Living systems feed on molecules rich in free energy and use the energy to do work. The final products are molecules poor in free energy (see Figure 1.7). To obtain energy-rich molecules in the first place, it is necessary to break out of the isothermal condition and obtain work for synthesis using the heat engine of photosynthesis. In this way, a fraction of the energy flowing from the hot sun to the cooler earth is captured by photosynthetic pigments and transduced into chemical energy.

*Figure 1.7*

**The energy cycle of the biosphere**

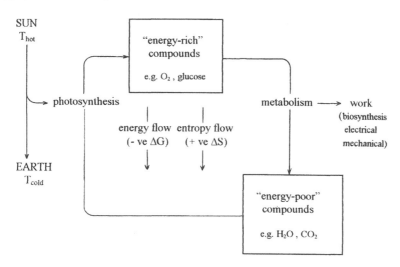

Organisms that use the energy from the oxidation of organic compounds (*heterotrophs*) are able to do work such as biosynthesis, ion gradient formation and mechanical movement by converting 'energy-rich' molecules such as carbohydrates and fats into 'energy-poor' molecules such as water and carbon dioxide. The overall direction of energy flow is determined by a drop in Gibbs free energy and an overall rise in entropy. Synthesis of the 'energy-rich' molecules by *phototrophs* involves the input of energy from the sun. (A few organisms (*chemolithotrophs*) can use the energy of chemical reactions involving simple surrounding materials such as hydrogen or hydrogen sulphide.)

## Summary

- Energy flow and an information system are necessary requirements for life. The nucleosides are molecules that bridge these two concepts. They are used to form RNA, the link between the information stored in DNA and the product of that information, proteins. As nucleoside triphosphates, they link energy-generating reactions in metabolism to energy-utilising reactions.

- Energy changes can be quantitated by the laws of thermodynamics. The first law states that energy can neither be created nor destroyed (this also applies to nuclear energy if we include the rest-mass energy of atoms). The second law states that any spontaneous change must be accompanied by an increase in the total entropy content of the universe. Entropy is a measure of the distribution of energy, high entropy being related to a more random distribution.

- It is possible to convert the random motion of molecules (heat) into work as long as a temperature difference exists between the system and surroundings. Living systems are isothermal, with no significant temperature differences within cells or organisms. Similarly, any temperature difference between organisms and their surroundings is too small to be used to supply the work energy for life. Living systems therefore cannot use heat generation to do work. Instead, they exist by converting the chemical energy of molecular bonding into useful forms of work such as the biosynthesis of macromolecules, the formation of ion gradients and the mechanical movement of their structures. The initial bonding energy is formed by photosynthesis, itself driven by the temperature difference (radiation) between the sun and the earth.

## Selected reading

Atkins, P.W., 1995, *The Second Law*, 2nd edn, New York: Scientific American Books, W.H. Freeman. (An excellent and popular treatment of the second law of thermodynamics)

Harold, F.M., 1986, *The Vital Force: A Study of Bioenergetics*, New York: W.H. Freeman. (One of the first textbooks to cover the full range of bioenergetics at the undergraduate level)

Monod, J., 1972, *Chance and Necessity*, Glasgow: Collins. (Emphasises the key concept of specificity for the creation of order)

Morowitz, H.J., 1968, *Energy Flow in Biology*, New York and London: Academic Press. (An early quantitative treatment of the subject)

Prigogine, I. and Stengers, I., 1984, *Order Out of Chaos*, Toronto, New York, London, Sydney: Bantam Books. (A readable account of irreversibility and time)

Schrödinger, E., 1945, *What is Life? The Physical Aspect of the Living Cell*, London and New York: Cambridge University Press. (Worthwhile for historical interest)

# 2 Energy Requirements and Energy Expenditure

## 2.1 Calorimetry

### 2.1.1 Direct calorimetry

The bomb calorimeter can be used to measure enthalpy changes when various molecules are oxidised to carbon dioxide and water. Table 2.1 shows that different food materials have different enthalpy values of oxidation when expressed in units of $kJ\,mol^{-1}$. Thus glucose has a value of $2817\,kJ\,mol^{-1}$, approximately twice that of lactate. Palmitic acid, a typical fat, has a value of $10\,040\,kJ\,mol^{-1}$, whereas a simple amino acid such as glycine has a value of $979$ $kJ\,mol^{-1}$. However, when expressed in units of $kJ\,g^{-1}$ a pattern emerges. Carbohydrate material gives a relatively con-

**Table 2.1** Energy yields from oxidation of various materials to carbon dioxide and water

| Substance | Energy yield | | | |
|---|---|---|---|---|
| | | | 'calorific value' | |
| | $kJ\,mol^{-1}$ | $kJ\,g^{-1}$ | $kcal\,g^{-1}$ | $kcal\,g^{-1}$ wet weight[a] |
| glucose[b] | 2 817 | 15.6 | 3.7 | – |
| lactate[b] | 1 364 | 15.2 | 3.6 | – |
| palmitic acid[b] | 10 040 | 39.2 | 9.4 | – |
| glycine[b] | 979 | 13.1 | 3.1 | – |
| carbohydrate[c] | – | 16 | 3.8 | 1.5 |
| fat[c] | – | 37 | 8.8 | 8.8 |
| protein[c] | – | 17 | 4.1 | 1.5 |
| ethyl alcohol[c] | – | 29 | 6.9 | – |
| lignin | – | 26 | 6.2 | – |
| coal | – | 28 | 6.7 | – |
| oil | – | 48 | 11 | – |

[a]Weight plus associated water.
[b]Values equivalent to $-\Delta H$.
[c]Average values used for calculation of metabolisable energy of foods.

stant value of around $16\,kJ\,g^{-1}$, close to the value for amino acids and proteins. In contrast, most fats give values around $37\,kJ\,g^{-1}$. These values are given relative to the dry weight of the material. In fact carbohydrate and protein, but not fat, in the cell, are usually associated with bound water. The values relative to wet weight for these two materials are therefore lower (Table 2.1).

For historical reasons, the energy change per gram is often expressed in terms of calories rather than joules. The value of $3.8\,kcal\,g^{-1}$ is termed the *calorific value* for carbohydrate. Fats give a higher calorific value, around $9\,kcal\,g^{-1}$, which is not surprising considering that the carbon atoms in fat are generally more reduced than in carbohydrate. Fats are therefore a more condensed form of energy storage than carbohydrate. The calorific value of amino acids varies according to the nature of the side-chain. It is interesting to compare the calorific values of various foods with the calorific values for coal and oil. These materials comprise compressed and highly reduced organic matter and, as might be expected have higher calorific values, around $7\,kcal\,g^{-1}$ and $11\,kcal\,g^{-1}$, respectively. Lignin, the main starting material for coal and oil from plants, has an average value around $6\,kcal\,g^{-1}$.

The first law of thermodynamics is very useful in calorimetry. Effectively it means that the energy released when burning food molecules in the laboratory is exactly the same as when the same food molecules are oxidised to carbon dioxide and water by the chain of biochemical reactions in the body. In the former case all the energy is released as heat (if the reaction is carried out at constant pressure), whereas in the latter some of the energy is also used to do work such as biosynthesis, ion gradient formation and mechanical movement. A useful trick is to realise that, over a reasonable period of time, most of the work energy in living systems will eventually dissipate as heat. Energy transduced into biosynthesis of new tissue will be matched by the energy released during catabolism of old tissue. Without growth, mature organisms will match energy intake with energy expenditure and the rate of heat output can be taken as a measure of the rate of energy expenditure. For growing organisms, an additional amount can be added, equivalent to the energy requirement for the net biosynthesis of new tissue (see Box 2.1). Hence calorimetry provides us with a very convenient method for measuring the steady-state energy utilisation of living systems.

Several calorimeters have been devised for working with large animals, including humans. These directly measure the

Box 2.1 **Energy intake and energy expenditure in humans**

At steady state, the energy input (food) will match energy output (work and eventually heat). Any imbalance will lead to weight gain or loss.

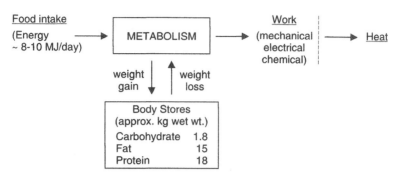

The stores of carbohydrate, fat and protein available for metabolism would last approximately 2–3 months on starvation before progressive muscle weakening caused death. A daily intake of 10 MJ is equivalent to 1.6 kg carbohydrate (wet weight) or 0.27 kg fat or 1.6 kg protein (wet weight) (see Table 2.1). A weight loss greater than 0.27 kg per day in the early stages of a 'starvation diet' usually indicates most of the total weight loss is due to loss of carbohydrate with associated water. An equally large gain will occur on stopping the diet. The aim of most diets is to lose fat. This should be achieved by the gradual introduction of a low-calorie diet and a steady weight loss of not more than 0.1 kg per day.

heat given out by the animal over a period of time. Some, such as the Atwater–Rosa calorimeter, comprise a large insulated chamber with heat production being measured by the temperature difference between inflow and outflow of a cooling stream of water piped through the chamber containing the subject (usually human). Others are more truly adiabatic, where the heat produced raises the temperature of the surrounding walls. Thermocouples monitor the temperature rise of the surrounding material and it is a relatively simple calculation to relate this to the heat changes in the chamber. A portable suit calorimeter has even been developed where the subject (again usually human) wears a water-cooled undergarment covered by an insulating outer layer.

### 2.1.2 *Indirect calorimetry*

Although direct calorimetry has its uses, the apparatus needed is often very cumbersome and expensive. It is possible to estimate energy output in an alternative, indirect, manner. This depends on the quantitative measurement of oxygen consumed and carbon dioxide produced during metabolism. The oxidation of 1 mole of glucose to carbon dioxide and water, according to the equation

$$C_6H_{12}O_6 + 6O_2 = 6CO_2 + 6H_2O$$

will consume $6 \times 22.4$ litres (6 moles) of oxygen, produce $6 \times 22.4$ litres of carbon dioxide and yield 2817 kJ. The ratio of heat evolved to oxygen consumed (or carbon dioxide produced) is $2817/(6 \times 22.4)$ which is $21 \, kJ \, l^{-1}$. Hence, if we can measure the quantities of oxygen or carbon dioxide, we have a simple means of calculating the quantity of heat evolved.

The calculations for fat and protein are slightly more complicated, but the same principles hold. Taking palmitic acid as the simplest example, we have

$$C_{15}H_{31}COOH + 23O_2 = 16CO_2 + 16H_2O$$

which yields $10040 \, kJ \, mol^{-1}$. In this case, the ratio of heat evolved to oxygen consumed is $19.5 \, kJ \, l^{-1}$. Other fatty acids will vary in their heat of combustion depending on the degree of unsaturation and whether they are esterified as triglyceride, but values around $19$–$20 \, kJ \, l^{-1}$ are still the norm. Protein, on the other hand, cannot be fully oxidised into carbon dioxide and water because of the nitrogen present. This is excreted as urea or ammonia, depending on the species, and some measure of nitrogen excretion is required in order to calculate an accurate value for the heat evolved. In humans, 1 g of urinary nitrogen arises from the metabolism of around 6.3 g of protein. Suprisingly, when this is taken into account, a value of around $20 \, kJ \, l^{-1}$ of oxygen is again found. It seems that the 'energy equivalence of oxygen' is approximately the same whatever the mixture of carbohydrate, fat or protein in the food.

Variation does occur in the ratio of carbon dioxide produced to the oxygen consumed for different foods. This ratio is known as the *respiratory quotient* (RQ). For glucose and most carbohydrate material the value is 1.0. For palmitic acid, in the example above, the RQ can easily be seen to be 0.7, which turns out to be typical for most fats. Values for amino acids vary according to the nature of the side-chain but an average value of 0.81 is usually taken for protein. The astute reader will now see that, if excretory nitrogen is measured together with the quantities of carbon dioxide and oxygen, the relative components metabolised of a mixture of carbohydrate, fat and protein can be calculated (see Box 2.2).

Average heat evolved per litre of oxygen consumed by living organisms is 20 kJ whatever the fuel.

*Box 2.2*  **Indirect calorimetry—an example**

**Problem**

Calculate the relative contributions of protein, fat and carbohydrate to the energy expenditure of a human given the following measurements of oxygen consumption, carbon dioxide evolution and urinary nitrogen excretion of a subject over a period of 5 hours.

| | |
|---|---|
| Litres of oxygen consumed | 71.6 |
| Litres of carbon dioxide produced | 60.4 |
| Urinary nitrogen | 1.6 g |

**Answer**

*Total heat production*
Since the heat production for humans is approximately $20 \, kJ \, l^{-1}$ of oxygen consumed, whatever the fuel, then the total heat production over the 5 h period is $20 \times 71.6 = 1432 \, kJ$. This is equivalent to $80 \, J \, s^{-1}$ (or 80 W).

*Protein*
Assuming 1 g of nitrogen arises from the metabolism of 6.3 g of protein (see text), then the total protein metabolised is 10.1 g. Using a value of $17 \, kJ \, g^{-1}$ for the oxidation of protein (see Table 2.1), then the contribution of pro-

tein oxidation to the total heat production is 172 kJ. This would involve the consumption of 8.6 litres of oxygen and the production of 7.0 litres of carbon dioxide (RQ for protein = 0.81).

*Carbohydrate and fat*
Subtracting the volumes of oxygen and carbon dioxide involved in protein metabolism leaves 63 litres of oxygen consumed and 53.4 litres of carbon dioxide evolved owing to the metabolism of non-protein material. If we say that $x$ litres of the oxygen are used for fat oxidation, then the amount of carbon dioxide formed from fat would be $0.71x$ litres. The remaining carbon dioxide $(63 - x)$ would be produced from carbohydrate (remember that the RQ for carbohydrate is 1). Since the total carbon dioxide from these two sources is 53.4 litres, we have,

$$0.71x + (63 - x) = 53.4$$

Hence $x = 33.1$ litres. We can now easily calculate the relative contributions of fat and carbohydrate to the heat production as 662 kJ and 598 kJ, respectively. This heat production arises from the oxidation of 17.9 g of fat and 37.4 g of carbohydrate (see Table 2.1).

# 2.2 Energy expenditure

There have been many measurements of the energy expenditure of living systems. An example for humans is given in Table 2.2, where indirect calorimetry has been used to provide average measurements of energy output during different activities in day-to-day life. It is important to note that even at rest a substantial energy requirement is needed. This is the *basal (or resting) metabolic rate*, defined as the rate of heat production under conditions where the subject is at complete rest with no physical work being carried out. Since energy is required for temperature regulation, the subject being measured should be maintained at a thermoneutral temperature. Food absorption and digestion should also be absent since this can increase metabolism by as much as 30%, an effect previously known as the specific dynamic action of food but now more generally termed the *thermogenic effect of food*. The basal metabolic rate for humans is around 70 W (equivalent to $70 \times 60 \times 60 \times 24 = 6048 \, kJ/$day). The value is slightly higher for men than for women

**Table 2.2**  Typical values for daily energy expenditure in humans

| Activity | Time (min) | Energy cost ($kJ\ min^{-1}$) | Total energy expenditure (kJ) |
|----------|-----------|----------|----------|
| lying | 540 | 5.0 | 2700 |
| sitting | 600 | 5.9 | 3540 |
| standing | 150 | 8.0 | 1200 |
| walking | 150 | 13.4 | 2010 |
| TOTAL | 1400 | – | 9450 |

The values were recalculated from measurements derived by indirect calorimetry of students by Haslam and Banner (1991, *Biochem. Soc. Trans.* **19**, 433S). A value of around 10 000 kJ is usually taken to be the daily energy requirement for humans, of which 6000 kJ will be the contribution by the basal metabolic rate. The calorific equivalent of the daily energy requirement will be 2400 kcal. (Note that the latter units are kilocalories not calories. How many times have you seen a diet in the press recommending daily requirements in calories, with a small 'c'? The dieter would soon starve to death.)

and can be up to a third higher in growing young children. The basal metabolic rate is also higher, by around 10%, than the sleeping metabolic rate because of the additional energy expenditure of wakefulness.

### 2.2.1  *Energy expenditure and size*

Many attempts have been made to relate the basal metabolic rate to body size. For aerobic organisms the relationship seems to be proportional to body mass raised to the power 0.75:

$$\text{basal metabolic rate} = a \cdot M^{0.75}$$

The constant of proportionality $a$ varies with phylum and even class, although little change is seen in the value of the exponent. For example, the basal metabolic rates for mammals are higher than for cold-blooded vertebrates such as reptiles when compared at the same body mass, but the slopes of the lines relating metabolic rate with mass are similar (see Figure 2.1). Even the metabolic rates for trees and microorganisms fall on lines with similar slopes but with different values of $a$. The results suggest some common mechanism to explain the rate–mass relationship.

The separate contributions of different metabolic reactions to steady-state heat production are not yet known. Brand (1990) has proposed that mitochondrial activity is important to the basal metabolic rate in aerobic organisms. He has related the total mitochondrial inner membrane area to the

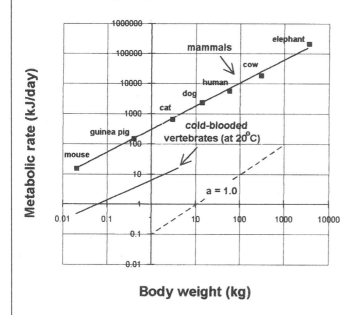

Figure 2.1
**Daily metabolic rates of various organisms
in relation to body weight**

Note that both scales are logarithmic. The slopes of the lines for aerobic organisms are the same ($a \approx 0.75$) although the position of the regression lines are different for warm-blooded as opposed to cold-blooded animals. (Plotted using data from Kleiber (1947) and Schmidt-Nielsen, K., 1970, *Fedn. Proc.* **29**, 1524–1532.)

basal metabolic rate over a wide range of body masses in both warm and cold blooded animals. Mitochondria conserve the energy of oxidation of metabolites such as pyruvate by the creation of a gradient of protons across the inner mitochondrial membrane. The gradient can then be used to catalyse the movement of other ions across the membrane and also for the synthesis of ATP (see Chapter 7). The free energy stored in the gradient can be lost if the protons simply leak back across the inner membrane. Brand suggests that a significant contribution to heat production by mitochondria is the leak of protons across the mitochondrial inner membrane. In fact, the total number of liver mitochondria in mammals has been shown to be proportional to $M^{0.72}$. As might be expected from this argument, anaerobic organisms have a relationship more directly proportional to body mass.

The ability to store food (for example carbohydrate or triglyceride) is generally proportional to the mass, $M$, of an organism. Since energy expenditure relates to $M^{0.75}$, problems of size can arise. A large animal such as a camel has plenty of body size to store food for its energy needs. As size decreases, the difference between the two mass functions, food storage and energy expenditure, converges (Figure 2.2). A mouse has relatively smaller energy stores to meet

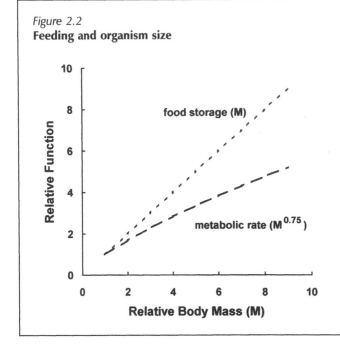

*Figure 2.2*
**Feeding and organism size**

The amount of food stored by an organism depends on its body mass $M$. Metabolic rate, however, depends on $M^{0.75}$. As body mass increases, the organism is able to store larger amounts of food to sustain its energy demands and can go longer between meals. As body mass decreases, the difference between food storage and energy demand converges and small organisms have to eat more frequently.

its energy demands. Small animals therefore have to spend correspondingly longer times foraging for food than larger animals: they have to eat more frequently.

## 2.2.2 *Physical activity*

The energy cost of physical activity will depend on the nature of the exercise and will be very variable. The maximum value achieved at steady state is probably limited by the efficiency of oxygen delivery to the muscles. The maximum oxygen consumption during exercise in mammals rarely exceeds ten times that during rest. Incidentally, it also shows the same relationship with body mass as basal metabolic rate. Blood flow increases dramatically during exercise, but there is still insufficient oxygen to support the theoretical maximum power output of muscle. Muscle contraction requires ATP. The action of *creatine kinase* helps to maintain ATP levels by phosphorylating ADP at the expense of *creatine phosphate*, a short-term store of phosphorylated compound (arginine phosphate is used in some invertebrates instead of creatine phosphate):

The rate of oxygen delivery to muscles usually limits the maximum energy expenditure in mammals.

creatine phosphate + ADP = creatine + ATP          (2.1)

**creatine phosphate**

Sudden bursts of activity can also be supported by increasing the activity of glycolysis (misnamed 'anaerobic' metabolism). The amount of ATP generated in glycolysis by the oxidation of glucose to pyruvate is less than one-tenth of that generated by the oxidation of pyruvate to carbon dioxide and water in mitochondria (see Chapter 5). Nevertheless, the reactions of glycolysis do not depend on oxygen and can be increased rapidly in response to hormonal signals. In mammals, the catecholamine hormones, epinephrine (adrenaline) and norepinephrine (noradrenaline), are released during stress or strenuous exercise (the so-called 'fight or flight' response) and lead to the mobilisation of glycogen and triglyceride as well as dilating blood vessels to increase blood flow. During intense exercise, the flux rate through glycolysis in muscle exceeds that through the citric acid cycle. The animal is rapidly provided with a source of ATP to either fight or escape. Over a few tens of seconds, this causes lactic acid to accumulate in the muscle, with a corresponding fall in pH as the acid dissociates to lactate plus $H^+$ (Figure 2.3). Unless cleared into the blood stream, the rise in lactate and $H^+$ (*lactate acidosis*) can rapidly lead to fatigue, the inability of the muscle to maintain power output. It is not known in detail how acidosis leads to fatigue. The rise in $H^+$

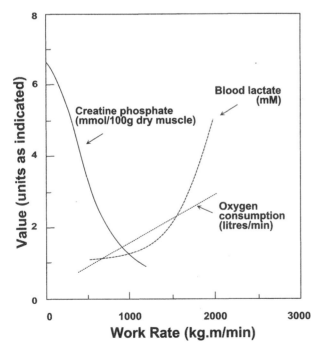

*Figure 2.3*
**Metabolite changes during exercise**

As the work rate increases, the oxygen consumption increases proportionately. Creatine phosphate levels in quadriceps muscle fall, but blood lactate does not increase until higher work loads when the rate at which lactate enters the blood exceeds its removal. (Replotted from data given in Hughes, E.F., Turner, S.C., and Brooks, G.A., 1982, *J. Appl. Physiol.* **52**, 1598–1607 and Bergström, J., 1967, *Physiology of Muscular Exercise*, C.B. Chapman, ed., New York: American Heart Association, pp. 191–196.)

concentration could inhibit glycolysis by affecting phospho-fructokinase, could interfere with muscle contraction by affecting $Ca^{2+}$ binding to troponin, and could stimulate pain receptors. Whatever the mechanism, the initial power increase, before fatigue comes into play, can support a trained athlete for over 100 metres. Longer distances depend more on mitochondrial activity and a supply of adequate fuels such as carbohydrate and triglyceride to match the power output needed (Table 2.3).

Lactate produced by active muscle diffuses into the blood and is carried to the liver, where it can be oxidised back to pyruvate. The pyruvate can then be converted to glucose by the process of *gluconeogenesis* (the synthesis of glucose from pyruvate or other precursor substrates such as alanine and glycerol) and the glucose returned to the bloodstream to be used again by muscle. Thus muscle and liver take part in a cycle of glucose and lactate, the *Cori cycle*. Muscle itself can consume lactate and, during prolonged exercise, lactate release declines as heart and skeletal muscle increase lactate consumption. Measurements of net release of lactate are liable to be underestimates of the total lactate production in muscle during exercise.

The rate of energy expenditure (power output) of an organism will vary according to activity. For example, flight imposes a major energy demand on birds. The demand is not constant. Initially, flight becomes more efficient with speed as the movement of air over the wings increases lift. As flight speed increases, however, the drag forces increase until it is no longer efficient to go any faster. For birds such as hummingbirds, there is also a large energy demand at zero speed to enable the bird to hover. Birds typically show a U-shaped curve of power output with flight speed (Figure 2.4). Tucker has pointed out that it is possible to express the energetic cost of movement, along a level path, by dividing the power or rate of energy expenditure, $dE/dt$ by the speed of movement, $dx/dt$ to give the ratio $dE/dx$. This ratio is the energy

**Table 2.3** Relative contribution of aerobic (mitochondrial) ATP production for athletic activity

| Activity | Contribution of mitochondrial metabolism to ATP production |
| --- | --- |
| 100 m sprint | <1% |
| 400 m run | 25% |
| 1500 m run | 65% |
| 5000 m run | 90% |
| marathon | >99% |

cost of movement per unit distance and is not a constant value (see Figure 2.4). There will always be some energy expenditure, even at rest, because of the necessary metabolic reactions of the body. At low speeds, the proportion of the total energy expenditure contributed by the basal metabolic rate will be high. The energy cost of movement will therefore be high (mathematically infinite at zero speed when $dx \to 0$). The cost of movement will fall as the speed increases but will rise again as speed becomes limited by the efficiency of the muscles. There will be a minimum value at which the animal can move with the least energy expenditure. Tucker (1975) gives the example of a human weighing 70 kg. The most efficient speed with regard to energy expenditure is around $1.8 \, \mathrm{m \, s^{-1}}$ (3.9 mph). At this speed, the energy expenditure is 450 W (Figure 2.5).

A further refinement is to divide the ratio $dE/dx$ by the body weight (expressed in terms of the gravitational force, 9.8 N, acting on body mass) in order to compare the energetic efficiencies $dE/dx$ of different sized animals. Figure 2.6 shows the results of this comparison for a variety of species. The costs of movement vary widely but there is an obvious grouping for the different methods of movement. Swimmers are the most efficient, followed by flyers and then runners. The size distribution for migratory animals suggests that a value of 2 or less is needed for sustained migratory move-

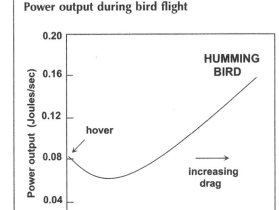

*Figure 2.4*
**Power output during bird flight**

As flight speed increases, the lift force generated by the increased flow of air over the wing increases. Power output then falls until drag resistance requires more power for flight. At zero speed, the hummingbird requires considerable power to hover. (Plotted from data given in Hainsworth (1981).)

*Box 2.3*   **Energy costs of movement—an example**

### Problem

A 50 kg woman has an energy expenditure of 7.8 kJ min$^{-1}$ when standing at rest. The extra energy cost of walking at 0.5 m s$^{-1}$ is 1.8 kJ min$^{-1}$ and this doubles for every 0.5 m s$^{-1}$ increase in speed. What is her most efficient speed of movement?

### Answer

The energy cost of movement can be calculated as the total energy expenditure divided by the walking speed and the body weight (mathematically this will be infinite at zero speed). The body weight is the product of the mass and the acceleration of gravity, in this case 50 × 9.8 N. A resting energy expenditure of 7.8 kJ min$^{-1}$ is equivalent to 130 J s$^{-1}$ (or 130 W). At 0.5 m s$^{-1}$ the extra energy demand is 30 W. The table below can be constructed to plot a graph of energy cost against speed.

The most efficient speed can then be seen to be around 1.5 m s$^{-1}$ or 3.3 mph.

| Speed (m s$^{-1}$) | Energy expenditure (W) | Cost of movement (J m$^{-1}$) |
|---|---|---|
| 0 | 130 | infinite |
| 0.5 | 160 | 0.65 |
| 1.0 | 190 | 0.39 |
| 1.5 | 250 | 0.34 |
| 2.0 | 370 | 0.38 |
| 2.5 | 620 | 0.51 |
| 3.0 | 960 | 0.65 |

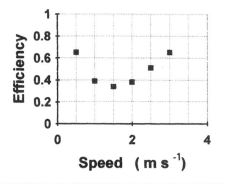

*Figure 2.5*
**The energy cost of walking for humans**

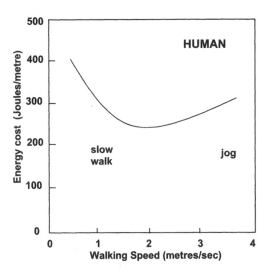

The cost efficiency of movement is obtained by dividing the rate of energy expenditure by the speed. This has a minimum value at which movement occurs with least energy expenditure. For a 70 kg human, this is around 1.8 m s$^{-1}$, or 3.8 mph. (Replotted from data given in Tucker, 1975.)

*Figure 2.6*
**Minimum costs of movement for a variety of animals**

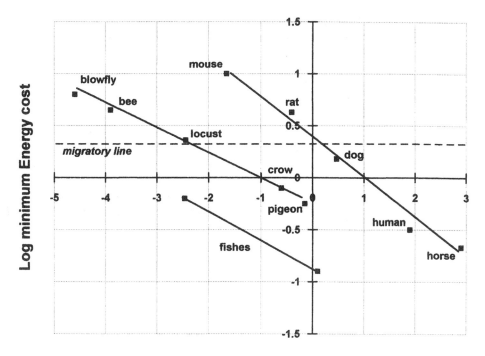

Although the cost of movement varies widely, animals tend to fall into groups based on their method of locomotion. The dotted line at a value of $dE/Wdx = 2$ indicates the 'migratory value', above which the energy cost of movement may be too great for the animal to sustain long distance migration. (Replotted from data in Tucker (1975).)

ment. Any greater value may require too much energy expenditure than the animal could afford (or store) for long distance migration.

## Summary

- Calorimetry can be used to measure the enthalpy values of food. Fats give a higher calorific value than either carbohydrate or protein. Steady-state calorimetry of living organisms can provide a measure of heat output with time which can be related to energy expenditure. When combined with measurements of oxygen consumption and nitrogen excretion, the relative contribution of carbohydrate fat and protein to steady-state metabolism can be calculated.

- Energy expenditure in mammals is the sum of the basal metabolic rate, the thermogenic effect of food, and physical work. The basal metabolic rate for aerobic organisms is proportional to body mass raised to the power 0.75. The constant of proportionality varies with phylum and class but the exponent is similar over a wide range of organisms. Anaerobes have a relationship more directly proportional to body mass.

- Physical activity requires changes in flux rate through metabolic pathways to provide sufficient ATP for energy demand. In mammals, high power output can be sustained over short periods by an increased rate of glycolysis but, over a few tens of seconds, the accumulation of lactate leads to acidosis and muscle fatigue. Sustained power output is limited by the rate of oxygen delivery to the muscles.

- The energy cost of movement per unit distance will pass through a minimum as speed increases. At low speed, the contribution by the resting metabolic rate will be high. At high speed, movement will become limited by the efficiency of the muscles. The most efficient speed with regard to energy expenditure will depend on the weight of the organism and the method of locomotion. Swimmers are the most efficient in movement followed by flyers and then runners.

# Selected reading

Brand, M.D., 1990, The contribution of the leak of protons across the mitochondrial inner membrane to standard metabolic rate, *J. Theor. Biol.* **145**, 267–286. (A reasoned attempt to relate physiological energy measurements to mechanisms at the molecular level)

Brooks, G.A., Fahey, T.D. and White, T.P., 1996, *Exercise Physiology: Human Bioenergetics and its Applications*, California, London, Toronto: Mayfield.

Brown, B.S., 1979, What does the kilojoule look like? *Biochem. Educ.* **7**, 88–89. (An amusing and instructive presentation of energy units)

Hainsworth, F.R., 1981, Energy regulation in hummingbirds, *Am. Sci.* **69**, 420–429.

Kleiber, M., 1947, Body size and metabolic rate, *Physiol. Rev.* **27**, 511–542. (A classic paper defining the basic problems of energy flow at the physiological level)

McLean, J.A. and Tobin, G., 1987, *Animal and Human Calorimetry*, Cambridge: Cambridge University Press.

Newsholme, E., Blomstrand, E and Ekblom, B., 1992, Physical and mental fatigue: metabolic mechanisms and importance of plasma amino acids, *Br. Med. Bull.* **48**, 477–495.

Tucker, V.A., 1975, The energetic cost of moving about. *Am. Sci.* **63**, 413–419.

## Study problems

1.  Calculate the heat energy released by burning a 10 g spoonful of sugar to carbon dioxide and water. (Use the data given in Table 2.1.)

2.  Calculate the nitrogen-free respiratory quotient for a starving human over a period of 24 h given a total oxygen consumption of 586 litres, a carbon dioxide production of 429 litres and a urinary nitrogen loss of 13.7 g. (Hint: First calculate the oxygen and carbon dioxide values associated with protein catabolism using the values given in Box 2.1. Then subtract these from the total gas values.)

3.  How many joules will be expended by a 50 kg woman climbing up 4 m of stairway?

4.  How long would you have to jog to 'burn off' a spoonful of sugar (10 g)? How long would it take to burn off a packet of peanuts (50 g, mainly fat)? (Hint: Refer to Table 2.1 and Figure 2.5.)

5.  How many metres of stairway could a 50 kg woman climb if all the energy available in metabolising a spoonful of sugar to carbon dioxide and water could be converted to work? Why is not all this energy available for work?

6.  A student spends around 16 h a day either sitting in lectures, studying or watching television. Walking between residence and college takes around 30 min of the day. The rest of the time is taken up in sleep. Calculate a daily energy requirement to satisfy his needs. Use the data given in Table 2.2. If this were to be met by carbohydrate alone, how many kilograms of carbohydrate would be needed? Another student takes a 1 h walk every morning and evening during the 16 h of awakeness. What extra energy demand is required to meet her needs?

# 3 Chemical Reactions and Energy Transduction

## 3.1 Free energy changes

In Chapter 1, we saw that the driving principle for any spontaneous change is the overall rise in entropy of the system *plus* surroundings. As long as the *net* change is positive, then it is perfectly possible for the entropy of the system to fall with a correspondingly larger rise in the entropy of the surroundings. We also saw that an alternative method for defining the direction of change in a system is the use of the Gibbs free energy function (see Equation (1.6)). The Gibbs free energy quantifies the maximum amount of energy change in a reaction that can be used to do work under isothermal conditions and at constant pressure, conditions appropriate to living systems. Three cases can be distinguished:

(i) Any spontaneous change has to occur in a direction for which $\Delta G < 0$.

(ii) A reaction for which $\Delta G > 0$ is thermodynamically forbidden. It will not take place unless coupled to another reaction which reverses the overall sign of $\Delta G$.

(iii) A system is at equilibrium when $\Delta G = 0$ for a change in any direction.

For a spontaneous change in a reaction A→B we conclude that the free energy change has to be negative. How can we quantify free energy changes experimentally?

We can start with a definition of the *Standard Free Energy Change*, $\Delta G°$ For a chemical reaction, this is the energy change when 1 mole of reactant A is converted to 1 mole of product B under standard conditions (molar concentrations and standard temperature and pressure). This is obviously a very difficult definition to use practically, since A and B would have to be maintained at molar concentrations throughout the reaction. Most reactions of interest to the biologist do not take place under standard conditions. In addition, the actual free energy changes will depend on the activities (concentrations) of A and B. During a reaction, the chemical potential ($\mu_i$) of each reacting species, $i$, will alter.

*Box 3.1*  **$\Delta G°$ and $\Delta G°'$**

Many biochemical reactions involve protons as reactants or products. Since the interior aqueous phase of most cells is at around pH 7, it would be inconvenient to have to use the concentration of protons in the calculations ($10^{-7}$ M at pH 7), so a convention has been accepted that the concentration, or activity, of protons is unity. Similarly, for reactions involving water, the activity of water is taken to be 1. (The concentration of pure water is 1000 $g\,l^{-1}$. Since the molecular mass of water is 18, this is an actual concentration of 55.6 M). Calculations of free energy where the activities of water and protons are taken to be 1 and do not change significantly during the reaction are indicated by the symbol $G'$. The standard free energy change for such reactions is then written as $\Delta G°'$.

The free energy change will be the sum of the changes in chemical potential multiplied by the respective number of moles $n_i$ undergoing the reaction:

$$\Delta G = \sum_i n_i \,\Delta\mu_i \tag{3.1}$$

For dilute solutions, the chemical potential of a solute species $i$ can be related to its concentration $c_i$ by the expression

$$\mu_i = \mu_i° + RT \ln c_i \tag{3.2}$$

where $\mu_i°$ is the chemical potential of species $i$ in a solution at standard concentration (1 M), $R$ is the gas constant ($8.314\,J\,mol^{-1}\,K^{-1}$) and $T$ is the temperature (usually taken to be 298 K for chemical reactions *in vitro*). From these relationships it can easily be shown that

$$\Delta G = \Delta G° + RT \ln \frac{[\text{products}]}{[\text{reactants}]} \tag{3.3}$$

For the reaction A→B, the concentration of products over the concentration of reactants is simply [B]/[A]. This is termed the *mass action ratio* for the reaction under those particular conditions.

Equation (3.3) can be looked at a little more deeply. When reactants and products are at equal concentrations, the logarithmic term is zero ($\ln 1 = 0$) and the actual free energy change is the same as the standard free energy change ($\Delta G = \Delta G°$). When the reactant concentrations are greater than the product concentrations, the logarithmic term will be a negative quantity (the logarithm of a number less than 1 is a negative number). The free energy change will therefore be more negative (more favourable) than the standard free energy change. Conversely, when the products are greater in concentration than the reactants, the free energy change

The free energy change of a reaction can be altered by changes in the concentrations (activities) of reactants and products.

will be more positive (less favourable). This is simply a quantitative statement of the *law of mass action*. The driving force of a reaction can be increased in a particular direction by appropriate changes in the concentrations of reactants and products (Figure 3.1).

At equilibrium, the free energy of the system will be at a minimum. There will be no free energy change for any infinitesimal change in concentration of reactants or products. Equation (3.3) will become

$$0 = \Delta G° + RT \ln \frac{[\text{products}]_{\text{equ}}}{[\text{reactants}]_{\text{equ}}}$$

But the ratio

$$\frac{[\text{products}]_{\text{equ}}}{[\text{reactants}]_{\text{equ}}}$$

is simply the *equilibrium constant* for the reaction, $K_{\text{equ}}$. Hence

$$\Delta G° = -RT \ln K_{\text{equ}} \qquad (3.4)$$

This formula gives us an easy and convenient method for calculating standard free energy changes. For example, we could mix some reactants and products together in a beaker with an appropriate catalyst and measure their concentrations when sufficient time had passed for the reaction to come to equilibrium. $\Delta G°$ then follows from a simple calculation.

*Figure 3.1*
**The free energy of a reaction mixture**

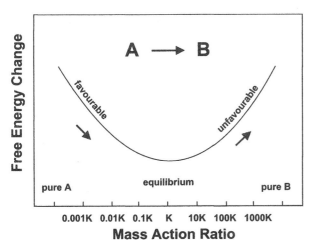

The Gibbs free energy of a particular reaction system will change as the concentration of reactants and products change, quantified by the mass action ratio. There will be a minimum value of free energy when the reaction is at equilibrium. The change in free energy ($\Delta G$) at any point is negative in the direction approaching equilibrium but positive moving away from the equilibrium position. At the equilibrium position, $\Delta G = -RT \ln K$ where $K$ is the equilibrium constant for the reaction.

It is important not to confuse $\Delta G°$ with $\Delta G$. The latter depends on the actual concentrations of reactants and products in the reaction mixture at any particular time and can be altered by altering those concentrations (see Box 3.2). $\Delta G°$ depends only on the equilibrium concentrations of the reactants and products. This gives a constant value for $\Delta G°$ and standard free energies for many reactions have been calculated and collected in useful tables.

## 3.2 Sequential reactions

In a sequence of metabolic reactions the overall change in free energy will be the sum of the free energy changes of the

---

Box 3.2   **Calculation examples**

The equilibrium constant for the reaction catalysed by phosphoglucomutase is 19:

glucose 1-phosphate $\rightleftharpoons$ glucose 6-phosphate

$$K_{equ} = 19$$

**Problem 1**

What is the value of $\Delta G°$ for this reaction?

**Answer**

Using Equation (3.4), we have

$\Delta G° = -8.314 \times 298 \times \ln 19$

$\quad = -7295 \, J \, mol^{-1}$

Note that $\Delta G°$ is negative for the equation as written, with glucose 6-phosphate as product. Under standard conditions, glucose-1-phosphate would be converted to glucose 6-phosphate.

**Problem 2**

What would be the value of $\Delta G$ if the concentrations of glucose 1-phosphate and glucose 6-phosphate were maintained at 10 mM and 1 mM, respectively?

**Answer**

Using Equation (3.3) with the value for $\Delta G°$ calculated above, we have

$\Delta G = -7295 + 8.314 \times 298 \times \ln(1/10)$

$\quad = -7295 - 5705$

$\quad = -13\,000 \, J \, mol^{-1}$

Note that the free energy change is even more negative (more favourable) than under standard conditions. The driving force for the reaction has been increased by increasing the concentration of the reactants relative to the products.

**Problem 3**

What would the be the value of $\Delta G$ if the concentrations of glucose 1-phosphate and glucose 6-phosphate were maintained at 0.01 mM and 1 mM, respectively?

**Answer**

Again, using Equation (3.3), we have

$\Delta G = -7295 + 8.314 \times 298 \times \ln(1/0.01)$

$\quad = -7295 - 11\,410$

$\quad = +4115 \, J \, mol^{-1}$

Note that in this case the free energy change is positive. The reaction will not spontaneously proceed to form glucose 6-phosphate. We have been able to *reverse the direction of the reaction by altering the concentration of reactants and products*. This is an important concept in metabolism.

individual reactions. This can easily be seen by considera-
tion of two sequential reactions and their mass action ratios:

(1)             A ⟶ B        with mass action ratio $K_1 = [B]/[A]$

(2)             B ⟶ C        with mass action ratio $K_2 = [C]/[B]$

Overall we have

(3)             A ⟶ C        with mass action ratio $K_3 = [C]/[A]$

Note that reaction (3) is the sum of reactions (1) and (2).
Compound B is the product of the first reaction but is consumed in the second. It acts as a *common intermediate* of the
two reactions. Also note that the mass action ratio for the
overall equation is the product of the mass action ratios for
reactions (1) and (2) ($K_3 = K_1 \cdot K_2$). Because of the logarithmic relationship between the mass action ratio and free
energy (Equation (3.3)), it follows that the free energy change
for the overall reaction is the sum of the free energy changes
for reactions (1) and (2) ($\Delta G_3 = \Delta G_1 + \Delta G_2$).

## 3.3 Coupled reactions

The logarithmic relation between $\Delta G°$ and the equilibrium
constant (Equation (3.4)) means that the direction of reactions with very large or very small values of $K_{equ}$ will be
very difficult to alter by a simple change of the mass action
ratio (i.e. using Equation (3.3)). Take, for example, an ender-
gonic reaction ($\Delta G > 0$) with a $\Delta G°$ of $+17\,100\,\mathrm{J\,mol^{-1}}$. This
corresponds to an equilibrium constant of 0.001, very much
in favour of the reactants rather than products. To alter the
mass action ratio sufficiently to produce a negative value for

**Table 3.1**   The variation of $K_{equ}$ with $\Delta G°$ at 25°C

| $K_{equ}$ | $\Delta G°$ (kJ mol$^{-1}$) |
|---|---|
| 0.001 | +17.1 |
| 0.01 | +11.4 |
| 0.1 | +5.7 |
| 1 | 0 |
| 10 | −5.7 |
| 100 | −11.4 |
| 1000 | −17.1 |
| 10 000 | −22.8 |

$\Delta G$, using Equation (3.3), the ratio of products over reactants would have to be greater than 1000 (see Table 3.1). Some examples of large mass action ratios are found in metabolism, but in general such large changes do not happen easily.

Metabolism therefore has a problem, if we can be allowed to think about it in those terms. Many reactions in the cell have, at first sight, a very unfavourable $\Delta G°$. For the reaction to proceed, the actual free energy change, $\Delta G$, has to be altered sufficiently to overcome the positive value of $\Delta G°$. The previous analysis has shown that if an endergonic reaction can be coupled to an exergonic reaction ($\Delta G < 0$) with a more negative free energy change, then the overall free energy change will also be negative. The examples shown were for simple sequential reactions where the product of one reaction acts as the reactant for the next. But what about a reaction such as the phosphorylation of glucose, a very common reaction in living systems? In mammals the absoption of blood glucose into the cell is accompanied by glucose phosphorylation. The negative charge on glucose 6-phosphate then prevents the molecule from diffusing back out of the cell and the glucose is 'activated' ready for further reactions of metabolism. However, written as

glucose + phosphate $\rightarrow$ glucose 6-phosphate + $H_2O$

gives a reaction which is very unfavourable under standard conditions. The $\Delta G°'$ value is $+13.8\,\text{kJ}\,\text{mol}^{-1}$ (see Box 3.1 for the distinction between $\Delta G°$ and $\Delta G°'$). To drive the reaction in the direction of formation of glucose 6-phosphate would require the blood glucose concentration to be greater than 100 mM and the intracellular glucose 6-phosphate concentration to be less than a few hundred micromolar. In practice, the reaction is coupled to the exergonic reaction of ATP hydrolysis.

ATP + $H_2O$ $\rightarrow$ ADP + phosphate     ($\Delta G°' = -31\,\text{kJ}\,\text{mol}^{-1}$)

Thus, overall, the free energies can be summed and the reaction can be written as

glucose + ATP $\rightarrow$ glucose 6-phosphate + ADP

$$(\Delta G°' = -17.2\,\text{kJ}\,\text{mol}^{-1})$$

The important question remains however, how are the two reactions coupled? Looking at the first reaction, the addition of phosphate to glucose, it can be seen that if the activity of inorganic phosphate could be increased sufficiently, together with a decrease in the activity of water, then the reaction

*Box 3.3*  **A note about equations**

Chemical equations are written so as to balance all atoms and charges. This is not the case with biochemical equations, where $H^+$ and metal ions are often left out and the molecules can be represented by words. It is important to realise which form is being used in any particular example so as not to get confused. For example,

$$ATP + H_2O \rightleftharpoons ADP + phosphate$$

is a biochemical equation where ATP refers to an equilibrium mixture of $ATP^{4-}$, $HATP^{3-}$, $H_2ATP^{2-}$, $MgATP^{2-}$, $MgHATP^-$, and $Mg_2ATP$ at specified values of pH and $Mg^{2+}$ concentration. A corresponding chemical equation would be

$$C_{10}H_{12}O_{13}N_5P_3^{4-} + H_2O \rightleftharpoons$$
$$C_{10}H_{12}O_{10}N_5P_2^{3-} + HPO_4^{2-} + H^+$$

although this is not unique since an alteration of pH or the addition of $Mg^{2+}$ would alter the way in which the equation has to be written. Sometimes equations are written in a mixed form which, apart from offending the purists, can lead to confusion. For example, $NAD^+$ is often used to represent the oxidised form of nicotinamide–adenine dinucleotide, whereas in fact the molecule has a net negative charge and quite reasonably could be written as $NAD^-$. The reaction catalysed by alcohol dehydrogenase could be written as

$$CH_3CH_2OH + NAD^- \rightleftharpoons$$
$$CH_3CHO + NADH^{2-} + H^+$$

or as

$$ethanol + NAD_{ox} \rightleftharpoons acetaldehyde + NAD_{red}$$

Both types of equation give useful information and the one to be used should depend on what emphasis is needed for any particular case. In the present volume, reactions will be mainly represented by biochemical equations, but some mixed forms will be used where the balance of hydrogen ions and charge has to be shown.

would be pulled to the right. Lowering the activity of water in the ATP hydrolysis reaction also has an important function. It inhibits the donation of the phosphoryl group of ATP to water. If this occurred, the energy of the reaction would not be conserved to drive the phosphorylation of glucose but would simply be released as the free energy of ATP hydrolysis. No coupling of the two reactions would take place even though the thermodynamics of the overall reaction is favourable. Mechanism as well as thermodynamics is important in life, and mechanisms of coupling require specific direction for the pathways of reaction. (Remember that random motion is associated with high entropy and a loss of free energy. For free energy to be conserved, vectorial movement has to be conserved.) The enzyme hexokinase achieves this feat. When glucose binds to the enzyme, a conformational change takes place which results in a 50-fold increase in the affinity of the active site for ATP. Two protein domains form a cleft for substrate binding and these rotate towards each other on glucose binding and partially close the binding site (Figure 3.2). ATP can then bind to a neighbouring site with high affinity. The consequences of the conformational change are that water is excluded from the active site and the catalytic groups of the enzyme are brought into proper alignment for phosphoryl transfer to take place (vectorial movement).

*Figure 3.2*
**The conformational change in hexokinase on glucose binding**

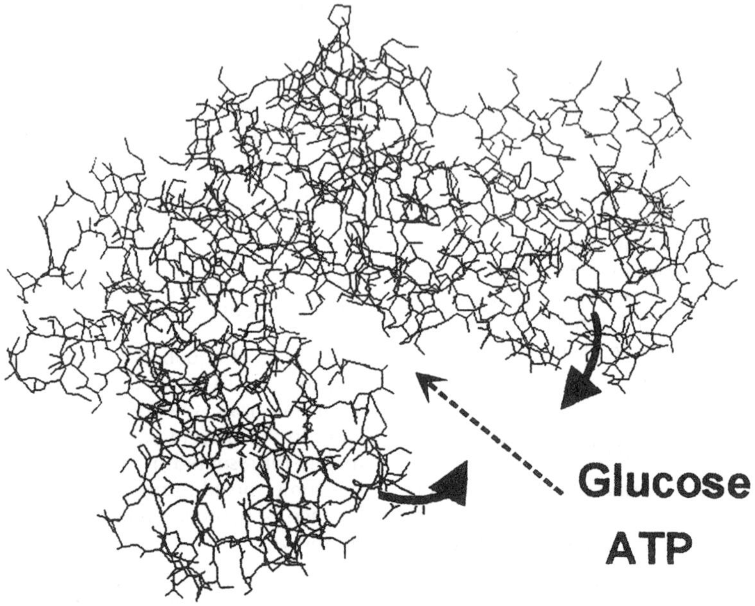

Binding of glucose to hexokinase induces rotation of two domains, forming a cleft around the active site. The conformational change increases the binding affinity of ATP and

*The lowering of water activity is crucial to the energetics of the dehydration reaction.* At an activity equivalent to $55\,M$, water could easily act as an acceptor of the phosphoryl group of ATP. Lowering the water activity favours donation to an alternative acceptor group on the protein and subsequently to the hydroxyl group in the 6-position of glucose. Direct measurements of water removal from hexokinase give a value of approximately 60–65 water molecules that leave the protein on glucose binding, of which half come from the region of the active site.

## 3.4 Binding energy

The example of hexokinase illustrates a general principle in the energetics of biochemical reactions. If the activity of a reactant or product can be altered relative to the activity of the other components of the reaction, then the equilibrium of the reaction will be shifted. A tighter binding of reactants relative to products (and vice versa) changes the activity of

Differential binding of reactants and products to a surface shifts the reaction equilibrium.

the reactant relative to the product. If the reaction is taking place on the binding surface, this will favour conversion into product. If the reaction is taking place in the bulk solution, it will make the reaction less favourable (Figure 3.3). The free energy of the reaction will alter according to Equation (3.3). A ten times tighter binding of a reactant *relative to product* will shift the free energy of the reaction on the surface by $-5.7 \, kJ \, mol^{-1}$ (Table 3.1). The effects of 'binding energy' are found extensively in many biochemical reactions, the best-known example being the synthesis of ATP by the $F_0F_1$-ATP synthase (see Chapter 7).

### 3.4.1 *Ligand binding and conformational changes*

A second important effect seen in the hexokinase example is the coupling of ATP binding with a conformational change in the protein. Several systems have evolved to take advantage of conformational changes associated with ligand binding and have utilised these changes to do the work of mechanical movement (*mechano-chemical coupling*). The binding energy of ligands to proteins is usually small and insufficient in itself to generate much mechanical work, although significant conformational changes can take place within the protein (for example, see oxygen binding to haemoglobin in Section 6.1). Significant force generation requires the free energy associated with binding *plus* a chemical reaction, such as ATP binding followed by hydrolysis. The coupling of the chemical reaction to the mechanical movement requires specific interaction between the force generator and the displaced object. In turn, there has to be close coupling of the conformational change to the binding of

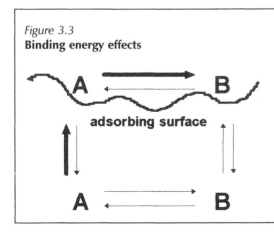

*Figure 3.3*
**Binding energy effects**

adsorbing surface

Binding of products or reactants to a solid substrate with different binding affinities alters the activities of the reactants and products and hence the overall free energy of the reaction. A tighter binding of A relative to B will shift the equilibrium of the reaction in the bulk solution in favour of A. The opposite will occur on the solid substrate since the relative activity of A on the surface is raised compared to B.

reactants and products. The best-known example of mechano-chemical coupling is muscle contraction (Box 3.4).

### 3.4.2 *Mechano-chemical coupling in actomyosin*

*Myosin* is a force-generating ATPase. It is a high-molecular-mass protein (470 kDa), comprising two heavy and two light chains. The chains intertwine to form two globular heads attached to a long filamentous tail. It readily interacts with *actin*, another filamentous protein of lower molecular mass (42 kDa) to give *actomyosin*. The interaction is controlled by other small proteins and calcium ions. Myofibrils of actomyosin can be seen clearly in skeletal muscle (Figure 3.4) and appear as parallel striations as the myosin and actin overlap in bands. A relative movement between myosin and actin can be brought about by the addition of ATP to isolated actomyosin fibres. Essentially, ATP binds to the myosin head group which acts as an ATPase and catalyses the formation of ADP and phosphate. These remain tightly bound to the myosin. Some conformational changes occur in the head group when hydrolysis occurs, but it is the *release* of the ADP and phosphate which is associated with the power stroke of movement.

The general principles of the mechano-chemical coupling cycle have been worked out (Figure 3.5), although several details are still not clarified by experiment. ATP binds very rapidly (around $10^6 \, M^{-1} \, s^{-1}$) to the myosin head group in the crosslinked actomyosin. The binding induces a

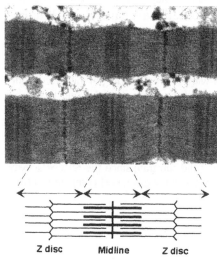

*Figure 3.4*
**Electron micrograph of muscle**

Z disc          Midline          Z disc

An electron micrograph of turkey breast muscle (×22,000) shows banded striations of actin (shown as thin lines in the accompanying diagram) and myosin (thick lines). The Z discs are structures that provide attachment points for the actin filaments and define the boundary between each contractile unit, or sarcomere. The actin and myosin filaments slide past one another during muscle contraction. (Courtesy of J. Pacy.)

conformational change which results in a tighter binding of the ATP molecule and a dissociation of the myosin head group from the actin filament. The link between the actin and myosin is now broken. The myosin ATPase catalyses the hydrolysis of the tightly bound ATP to produce tightly bound ADP and phosphate. Because the reactants and products remain bound, the equilibrium of the reaction is shifted to much lower values, conserving most of the free energy in the altered conformation of the protein. Isotope exchange experiments using $H_2^{18}O$ have shown that the oxygen isotope can be incorporated into more than one of the oxygen sites in the phosphate since the residence time on the protein is long enough for equilibration to produce multiple exchange of the phosphate oxygens. The conformation of the protein with bound ADP and phosphate is different from that of the protein with bound ATP, and re-association with actin takes place. The heads of the myosin chains form crossbridges with the actin filaments. However, the conformational energy is still insufficient to cause relative movement between the two proteins. The free energy of the reaction is only trans-

*Figure 3.5*
**Mechano-chemical coupling by myosin ATPase**

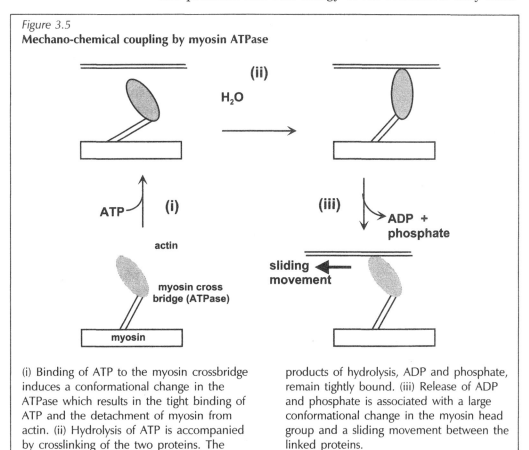

(i) Binding of ATP to the myosin crossbridge induces a conformational change in the ATPase which results in the tight binding of ATP and the detachment of myosin from actin. (ii) Hydrolysis of ATP is accompanied by crosslinking of the two proteins. The products of hydrolysis, ADP and phosphate, remain tightly bound. (iii) Release of ADP and phosphate is associated with a large conformational change in the myosin head group and a sliding movement between the linked proteins.

duced to mechanical work when the ADP and phosphate are released and the protein changes back to its original conformation. When this happens, the crossbridges alter their orientation, causing the attached actin to slide past the myosin head groups. The ATPase head groups act as force-generating crossbridges, linking the free energy from the chemical reaction of ATP hydrolysis to mechanical movement. The slow release of bound ADP and phosphate controls the rate of movement in steady-state turnover.

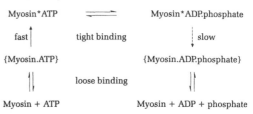

The basic principles of mechano-chemical coupling probably apply to other force-generating molecules, for example the interaction of *kinesin* and *dynein* with *microtubular protein* which is responsible for organelle movement in the cytosol.

## 3.5 Common intermediates

Most of bioenergetics at the molecular level is concerned with mechanisms to link exergonic reactions ($\Delta G < 0$) to endergonic reactions ($\Delta G > 0$). Some form of common intermediate is needed. In the case of the example of sequential reactions in metabolism, the common intermediate can simply be an intermediate metabolite. The product of the first reaction acts as a reactant of the second. Effectively the mass action ratios of the individual reactions are altered by the overall energetics to drive the metabolic pathway in a certain direction. If a particular reaction has a very unfavourable standard free energy change, it may be coupled to a separate exergonic reaction such as the hydrolysis of a phosphate ester (Box 3.4). The coupling intermediate can be an enzyme complex such as found for the hexokinase reaction or a protein conformational change in mechano-chemical coupling. There are two further coupling mechanisms found in living systems (Table 3.3). These are coupling using reduction/oxidation reactions (*redox coupling*), where electrons effectively act as common intermediates, and *chemiosmotic coupling*, where the exergonic and endergonic reactions are coupled through a shared gradient of molecules across a membrane. These will be dealt with in the following chapters.

*Box 3.4* **Phosphate esters as free energy intermediates**

The nucleoside triphosphates (NTPs) found in metabolism are essentially pyrophosphate ($P_2O_5$) molecules bound to the corresponding nucleotide (AMP, UMP, GMP, CMP). The chemical structure of ATP is shown in Figure 3.6. Like inorganic pyrophosphate, the phosphate esters on the nucleosides react with water with a release of energy. Thus the reaction

$$NTP + H_2O \rightarrow NDP + phosphate \qquad (3.5)$$

has a relatively large negative value of $\Delta G^{o\prime}$. The standard free energy change for the reaction, taking the activities of $H^+$ and water as unity ($\Delta G^{o\prime}$, see Box 3.1), is around $-31$ kJ $mol^{-1}$ depending on the concentration of any $Mg^{2+}$ ions present. Fortunately for biology, the nucleoside triphosphates are fairly stable in water even though the thermodynamics of the reaction favour hydrolysis. With appropriate enzymes therefore, the cell can use these molecules as chemical intermediates to drive condensation reactions in biosynthesis. Thus the synthesis of $X-Y$ from $X-OH$ and $H-Y$, represented by the (unfavourable) reaction

$$X-OH + H-Y \rightarrow$$
$$X-Y + H_2O \qquad \Delta G > 0 \qquad (3.6)$$

can be linked to reaction (3.5) to drive the overall reaction in a favourable direction:

$$X-OH + H-Y + NTP \rightarrow$$
$$X-Y + NDP + phosphate \quad (\Delta G < 0) \ (3.7)$$

The mechanism linking the two reactions depends on the NTP reacting with $X-OH$ to form the common intermediate $NDP-O-X$. It can easily be seen that it is essential in the linked reactions that water activity is lowered

as much as possible, otherwise the NTP will simply react with water ($H-OH$ instead of $X-OH$) and the free energy will be released as heat rather than being transduced into the bond energies of $X-Y$.

Other phosphate esters are also used as common intermediates, although not to the same extent as the nucleoside triphosphates. These include phosphoenolpyruvate, 3-phosphoglycerate, phosphocreatine and even pyrophosphate itself. They all differ in the standard free energy of hydrolysis (Table 3.2) because of their structural chemistry. For example glucose 1-phosphate has a slightly higher standard free energy of hydrolysis than glucose 6-phosphate because the phosphate is joined to an aldehyde group on the C-1 of the sugar rather than to an alcohol group which is found at the C-6 position.

**Table 3.2** Standard free energies of hydrolysis of some organophosphates

| Compound | $\Delta G^{o\prime}$ (kJ $mol^{-1}$) |
|---|---|
| phosphoenolpyruvate | $-62$ |
| phosphocreatine | $-43$ |
| ATP (and other NTPs) | $-31$ |
| ADP | $-28$ |
| pyrophosphate | $-28$ |
| glucose 1-phosphate | $-21$ |
| fructose 6-phosphate | $-16$ |
| AMP | $-14$ |
| glucose 6-phosphate | $-14$ |
| glycerol 3-phosphate | $-9$ |

*Figure 3.6*
**The structure of ATP**

The ATP molecule is effectively made from three building blocks: adenine, ribose, and a triphosphate unit.

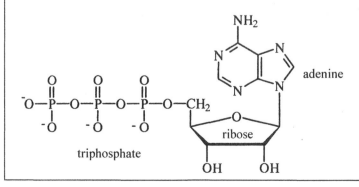

**Table 3.3**  Common coupling mechanisms found in living systems

| Coupling mechanism | Common intermediate | Example |
|---|---|---|
| sequential reaction | metabolite | glucose 1-P → glucose 6-P → fructose-6-P |
| chemical coupling | enzyme complex | glucose + ATP → glucose 6-P + ADP (hexokinase) |
| mechano-chemical | ligand–protein complex | ATP hydrolysis by actomyosin |
| redox reaction | electron | ethanol + $NAD_{ox}$ → acetaldehyde + $NAD_{red}$ |
| chemiosmotic | ion gradient | ATP synthesis coupled to electron transfer in photosynthesis and respiration |

*Box 3.5*  **ATP and the adenylate energy charge**

As discussed in Chapter 1 (Box 1.1) each nucleoside triphosphate seems to have evolved to be a common intermediate in a particular area of metabolism. ATP is mainly involved in electrical and mechanical work (ion pumps and contractile proteins), whereas UTP is involved in carbohydrate biosynthesis, GTP in protein synthesis and CTP in lipid synthesis. One reason why ATP is often referred to as the 'energy currency' of the body is that it is by far the most significant intermediate in cellular bioenergetics. Apart from periods of excessive growth, living systems expend most of their energy in the maintenace of ion gradients and the directed movement of molecules within the cell or across cellular membranes. Most of the energy demand of the human basal metabolic rate requires ATP turnover. It can be calculated that around 45 kg of ATP are involved per day. Since the total amount of ATP in the body at any one time is only around 1 g, then every second around $10^{20}$ molecules of ATP need to be formed from ADP and inorganic phosphate.

The enzyme *adenylate kinase* is present in most cells and catalyses the interconversion of ATP and AMP with ADP:

ATP + AMP = 2 ADP

Adenylate kinase is highly active and therefore the reaction will be in or near to equilibrium at all times. The relative concentrations will vary according to the amount of ATP present. Atkinson (1977) proposed a measure of the 'adenylate energy charge' of a system (EC). This is defined as

$$EC = \frac{ATP + 0.5\,ADP}{ATP + ADP + AMP}$$

The factor 0.5 is present because two molecules of ADP can generate one molecule of ATP. The value of EC will clearly range between 1.0 when all nucleotides exist as ATP and 0 when all exist as AMP. Any process that utilises ATP will lower the value. Conversely, any process generating ATP will raise the energy charge. Most healthy cells have values around 0.9–0.95. A drop to below 0.85 indicates onset of severe damage leading to cell death.

## Summary

- The direction of a chemical reaction depends on the sign of the free energy change. If this is negative then the reaction can proceed spontaneously, provided any activation energy can be overcome. The magnitude of the free energy change depends on the ratio of the concentrations (activities) of the products and reactants. Alterations in this ratio can alter the sign of the free energy change as well as its magnitude. Equations (3.3) and (3.4) provide important quantitative relationships for the application of thermodynamics to biology.

- In sequential reactions, the overall change in free energy is the sum of the free energies of the individual reactions.

- Unfavourable reactions, with a positive free energy change, can be coupled to more favourable reactions provided there is a common intermediate linking the two reactions. Coupling intermediates include metabolites, enzyme complexes, conformational changes in protein structure, electrons, and ion gradients across biological membranes.

## Selected reading

Atkinson, D.E., 1977, *Cellular Energy Metabolism and its Regulation*, New York: Academic Press. (Deals with the energetics of metabolic control)

Bagshaw, C.R., 1993, *Muscle Contraction*, 2nd edn, London: Chapman & Hall. (A clear explanation of the mechanism of muscle function)

Edsall, J.T. and Gutfreund, H., 1983, *Biothermodynamics: The Study of Biochemical Processes at Equilibrium*, Chichester: Wiley. (An advanced treatment of the application of thermodynamics to biochemical problems)

Klotz, I.M., 1978, *Energy Changes in Biochemical Reactions*, New York: Academic Press. (Plenty of quantitative examples)

Lehninger, A.L., Nelson, D.L. and Cox, M.M., 1993, *Principles of Biochemistry*, 2nd edn, New York: Worth. (Contains a clear treatment of bioenergetics applied to biochemistry. Lehninger was one of the masters in the field)

Nicklas, R.B., 1984, A quantitative comparison of cellular motile systems, *Cell Motil.* **4**, 1–5. (An interesting comparison of the power outputs of cellular and mechanical engines)

## Study problems

1. Consider the following reaction sequence:

$$\text{glucose 1-phosphate} \xrightarrow{[1]} \text{glucose 6-phosphate} \xrightarrow{[2]} \text{fructose 6-phosphate}$$

The equilibrium constant for reaction (1) is 19 and for reaction (2) is 0.52. Calculate:

(a) The equilibrium constant for the overall reaction of glucose 1-phosphate to fructose 6-phosphate.

(b) The standard free energy change $\Delta G°$ for the conversion of glucose 1-phosphate to fructose 6-phosphate.

2. Creatine kinase catalyses the phosphorylation of ATP in the reaction

$$\text{phosphocreatine} + \text{ADP} \rightarrow \text{ATP} + \text{creatine}$$

The standard free energies of hydrolysis for ATP (to ADP) and phosphocreatine (to creatine) are $-30.5\,\text{kJ mol}^{-1}$ and $-43\,\text{kJ mol}^{-1}$, respectively. The concentrations of various compounds in human muscle are as follows:

$$[\text{ATP}] = 10\,\text{mM}$$
$$[\text{ADP}] = 1\,\text{mM}$$
$$[\text{phosphocreatine}] = 30\,\text{mM}$$
$$[\text{creatine}] = 1\,\text{mM}$$

Calculate the change in free energy when the creatine kinase reaction occurs in human muscle at 37 °C.

3. Calculate the rate of ATP production (mmol s$^{-1}$) during a marathon run where the energy expenditure is 1 MJ h$^{-1}$. Assume $\Delta G'$ for ATP production in the cell is $-50\,\text{kJ mol}^{-1}$.

4. Discuss how it is possible to have a net flux through a reaction in a direction which has a positive free energy change.

5. Consider the reaction

$$\text{lactate} + \text{NAD}_{\text{ox}} \rightarrow \text{pyruvate} + \text{NAD}_{\text{red}}$$

This has a standard free energy change of $+27\,\text{kJ mol}^{-1}$. Will spontaneous formation of pyruvate occur by coupling the reaction to

(a) pyruvate $\rightarrow$ acetyl-CoA $\quad (\Delta G°' = -37.5\,\text{kJ mol}^{-1})$,

(b) phosphoenol pyruvate + ADP $\rightarrow$
$$\text{pyruvate} + \text{ATP} \ (\Delta G°' = -29.3\,\text{kJ mol}^{-1})?$$

6. The $\Delta G°'$ for the hydrolysis of sucrose to glucose plus fructose at 25 °C is $-23\,\text{kJ mol}^{-1}$. What is $\Delta G°$ for the same reaction?

# 4 Redox Reactions: Electrons as Common Intermediates

## 4.1 Redox couples and redox potentials

Many reactions in living systems involve either the reduction or the oxidation of a compound (*redox reactions*). Reduction involves the gain of electrons:

oxidised compound $(A_{ox}) + ne^- \rightarrow$

$$\text{reduced compound} \, (A_{red}) \qquad (4.1)$$

This may or may not be accompanied by the gain of protons:

$$A_{ox} + ne^- + mH^+ \rightarrow A_{red}H_m \qquad (4.2)$$

in which case $A_{red}$ is also acting as a proton acceptor. The numbers of electrons and protons do not have to match. Remember also that proton acceptance can take place without reduction, for example in the association of a carboxylic acid at low pH.

The reactions written above show the theoretical reduction reactions of a single redox couple and are known as *half reactions*. Electrons in aqueous medium are very reactive and have a short lifetime (around $10^{-6}$ s) before reacting with any nearby molecule. In practice therefore, redox reactions in the cell involve direct electron transfer from one redox couple to another. The electron can be said to be acting as a common intermediate linking the two couples. The electron donor is the *reductant* or *reducing agent* and the compound accepting the electrons is the *oxidant* or *oxidising agent*.

$$A_{red} \quad + \quad B_{ox} \rightarrow A_{ox} + B_{red} \qquad (4.3)$$

reductant    oxidant

The oxidised and reduced forms of A $(A_{ox}/A_{red})$ and B $(B_{ox}/B_{red})$ are known as *redox couples*. They need not necessarily have the same oxidation number. For example four reduced cytochrome $c$ molecules are required to reduce one molecule of oxygen to water:

$$4\text{cyt}\,c^{2+} + 4\text{H}^+ + \text{O}_2 \rightarrow 2\text{H}_2\text{O} + 4\text{cyt}\,c^{3+}$$

In the cell, various molecules act as electron carriers, helping to catalyse the transfer of electrons between other reacting metabolites (Figure 4.1). One of the most common is nicotinamide–adenine dinucleotide (a redox couple represented by $\text{NAD}_{\text{ox}}/\text{NAD}_{\text{red}}$). This water-soluble compound collects pairs of electrons released in metabolism, for example from glycolysis or the oxidation of fatty acids. It can diffuse in the aqueous compartments of the cell (but not cross cellular membranes) and transfer the electrons to various oxidants, for example pyruvate or oxygen. Other intermediate electron carriers in the cell can be fixed in protein assemblies, for example some of the cytochromes in the electron transfer chains of mitochondria and bacteria, or restricted to the apolar phase of phospholipid membranes, for example coenzyme Q (ubiquinone/ubiquinol).

The tendency of any particular redox couple to accept or donate electrons in a reaction depends on the *redox potential*

---

*Figure 4.1*
**Some intermediate electron carriers involved in metabolism (a–e)**

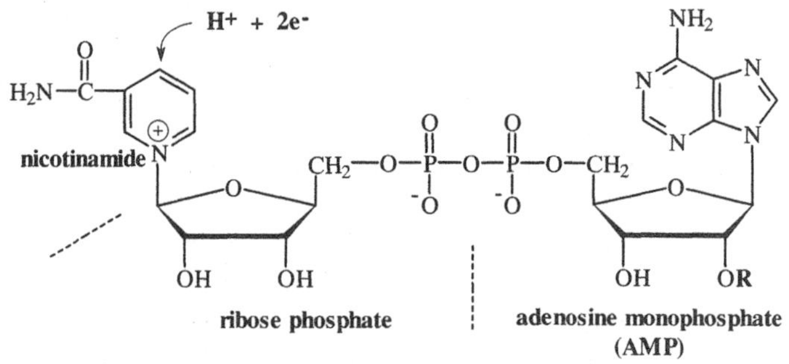

**(a) Nicotinamide adenine dinucleotide** (oxidised)

(a) Nicotinamide–adenine dinucleotides. The major soluble redox intermediates in metabolism are NAD and NADP. These differ by an additional phosphate group esterified at the C-2′ position on the adenylate ribose in NADP (R=H or R=phosphate). The reduced forms have two electrons and one proton more than the oxidised forms. $\text{NADP}_{\text{red}}$ mainly provides electrons for biosynthesis, whereas $\text{NAD}_{\text{red}}$ mainly functions to shuttle electrons to electron transfer chains.

*Figure 4.1*  **(continued)**

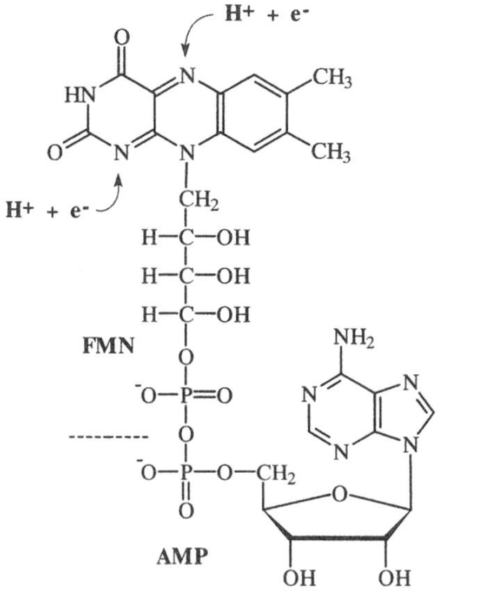

(b) Flavin–adenine dinucleotide. A protein-bound cofactor which accepts two electrons (with two protons) for full reduction. Unlike NAD, it can form a one-electron semiquinone intermediate. The molecule comprises adenosine monophosphate (AMP) linked to flavin mononucleotide (FMN).

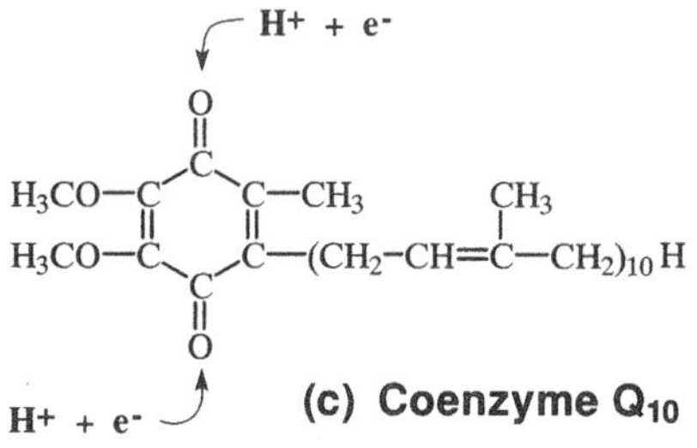

**(b) Flavin adenine dinucleotide** (oxidised)

**(c) Coenzyme Q$_{10}$**

(c) Coenzyme Q or ubiquinone. Coenzyme Q is a lipid-soluble redox intermediate that accepts two electrons (with two protons) for full reduction to ubiquinol. Like FAD and FMN, coenzyme Q can form a one-electron semiquinone intermediate. The length of the isoprenoid tail varies with species. The most common in mammals is 10 isoprenoid units long (CoQ$_{10}$).

*(continued)*

*Figure 4.1*   (**continued**)

## (d) Iron sulphur clusters

| simple iron centre | 2 iron/cluster | 4 iron/cluster |

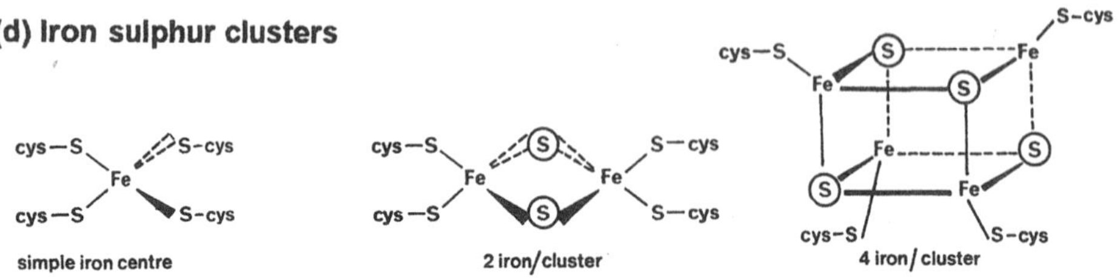

Rubredoxin type          Ferredoxin type

(d) Iron–sulphur clusters. These contain iron coordinated to sulphur, both inorganic (ringed sulphurs) and organic. The clusters are attached to protein and most cycle between an oxidised state and a one-electron reduced state.

## (e) Haem

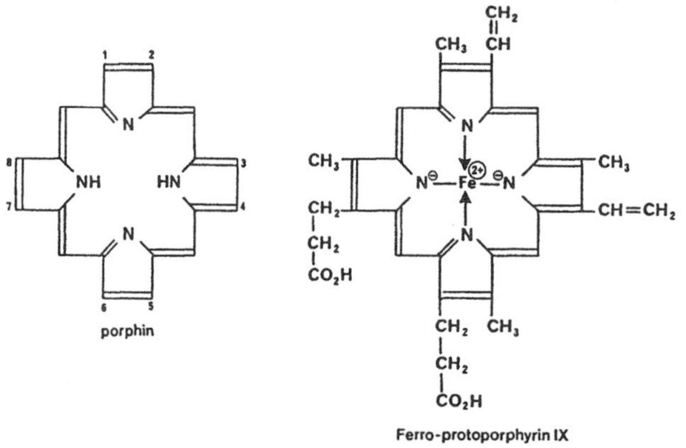

porphin

Ferro-protoporphyrin IX

| Haemprotein | position | group |
|---|---|---|
| haemoglobin<br>myogloblin<br>erythrocruorin<br>catalase<br>peroxidases (plant)<br>cytochromes class B | 1–8 as in protoporphyrin IX | |
| cytochromes class A | 2 — $CHOHCH_2CHCH_3(CH_2)_3CHCH_3(CH_2)_3CH(CH_3)_2$<br>8 — $CHO$ | |
| cytochromes class C | 2 — $CHCH_3S$–R–protein<br>4 — $CHCH_3S$–R–protein | |
| cytochromes class D | $2 - C{\overset{\text{R}}{\underset{\text{OH}}{\overset{H}{<}}}}$<br>7–8 saturated C—C bond | |

(e) Haem. Haem is a complex of iron coordinated to porphyrin which is in turn coordinated to protein, usually through a cystein residue represented by R. The side-chains on the porphyrin vary for different haem compounds. Addition of two hydrogen atoms at positions 7 and 8 in porphin gives the related structure found in chlorophylls, where magnesium (as $Mg^{2+}$) replaces iron. (From Wrigglesworth and Baum (1980).)

of the couple. A redox couple with a more negative potential will donate electrons to a redox couple of more positive potential (electrons are attracted to positive potentials). Under standard conditions, where all the components of the reaction are at molar concentrations (or 1 atmosphere for gases) the redox potential is represented as $E°$. When protons are involved, as shown in Equation (4.2), it is found to be convenient to regard the standard concentration of $H^+$ to be $10^{-7}$ M (i.e. pH 7) rather than 1 M. The standard redox potential is then written as $E°'$.

Just as the free energy of a reaction depends on the actual concentrations of reactants and products (see Equation (3.3)), the redox potential also depends on concentration. To see how the dependence comes about it is necessary to consider the *electrochemical potential* of the reacting species rather than the chemical potential. This takes into account the energy associated with an electric charge in a potential gradient. When a charge is moved in an electric field, the work done can be expressed as the product of charge and potential:

work done = charge × potential

The free energy change for a charged species transferred over a potential difference $\Delta E$ will be

$$\Delta G = -nF\,\Delta E \qquad (4.4)$$

**Michael Faraday** was the first person to quantitate the relation between chemical reactions and electricity. The Faraday constant is the factor that relates the energy involved in transferring charge between oxidising and reducing agents and the potential difference between the chemicals (Equation (4.4)).

where $n$ is the number of charges associated with the species and $F$ is the *Faraday constant* ($96\,485\,\mathrm{J\,V^{-1}\,mol^{-1}}$). The sign of $\Delta E$ is defined such that it has a positive value when the reaction is favourable ($\Delta G - $ ve). This happens when negative charge is attracted to a positive potential or positive charge is attracted to a negative potential. $\Delta E$ can be expressed as

$$\Delta E = -\frac{\Delta G}{nF} \qquad (4.5)$$

or, under standard conditions,

$$\Delta E° = -\frac{\Delta G°}{nF} \qquad (4.6)$$

Substituting for $\Delta G$ (using Equation (3.3)) gives

$$\Delta E = -\frac{\Delta G° + RT\ln\dfrac{[\text{products}]}{[\text{reactants}]}}{nF} \qquad (4.7)$$

or

$$\Delta E = \Delta E° - \frac{RT}{nF}\ln\frac{[\text{products}]}{[\text{reactants}]} \qquad (4.8)$$

Equation (4.8) is known as the *Nernst equation* and has wide application in biological systems. The equation for a single redox couple or half reaction is usually written in terms of $E$ rather than $\Delta E$ and the oxidised form is conventionally written on the left hand side (i.e. as reactant). This ensures that the sign of $\Delta E$ is positive when $\Delta G$ is negative. This convention can be confusing (compare for example different statements of the Nernst equation in textbooks of chemistry, physiology and biochemistry). However if we remember that $\ln(a/b)$ is the same as $-\ln(b/a)$ then the more usual form of Equation (4.8) for redox couples becomes

$$E = E^\circ + \frac{RT}{nF} \ln \frac{[\text{oxidised}]}{[\text{reduced}]} \tag{4.9}$$

Common-sense analysis of Equation (4.9) may help. When the oxidised form of the couple is at a larger concentration than the reduced form, then we would expect the actual redox potential to be more positive than the standard

**Box 4.1  A calculation example**

The standard redox potential $E^{\circ\prime}$ of the $NAD_{ox}/NAD_{red}$ couple is $-0.32\,V$.

**Question**

What is the actual redox potential, at 37°C, when the couple is 90% reduced?

**Answer**

The half reaction can be written as

$$NAD_{ox} + 2e^- + H^+ \rightarrow NAD_{red}$$

and involves the transfer of $2e^-$ ($n = 2$) and $1\,H^+$. The involvement of $H^+$ can be ignored since the question asks for $E^{\circ\prime}$ rather than $E^\circ$. From Equation (4.8), using a value for $T$ of 310 K, we have

$$E = -0.32 + \frac{8.314 \times 310}{2 \times 96\,485} \times \ln(10/90)$$
$$= -0.32 - 0.029$$
$$= -0.349\,V$$

*Hint:*
The quantity

$$\frac{RT}{nF} \ln \frac{[\text{oxidised}]}{[\text{reduced}]}$$

can be written in terms of logarithms to the base 10.

(remember that $\ln x = 2.3 \log_{10} x$). In that case $2.3RT/nF$ has a value of approximately 0.06 V for reactions where $n = 1$ (0.059 V at 25°C, and 0.61 V at 37°C). For two-electron reactions where $n = 2$, the value will be 0.03 V. Equation 4.9 can then be written as

$$E = E^\circ + 0.06 \log_{10} \frac{[\text{oxidised}]}{[\text{reduced}]} \tag{4.10}$$

Hence for every 10-fold change in the ratio of oxidised over reduced, the standard redox potential will change by 0.06 V for one-electron reactions and 0.03 V for two-electron reactions (see Table 4.1). The change will be in the positive direction when the oxidised concentration is increased and in the negative direction when the reduced concentration is increased.

In the example above, 90% reduced is approximately a ratio of 10. $\text{Log}_{10} = 1$, and since this is a two-electron reaction, the redox potential will be more negative than the standard potential by 0.03 V. The actual potential will therefore be $-0.35$ V.

*Box 4.2*  **A calculation example**

**Question**

The standard redox potential $E^{o\prime}$ for the pyruvate/lactate couple is −0.19 V. If a mixture of these two metabolites, both at 1 mM concentration, is added to a 1 mM solution of 10% reduced NAD ($E^{o\prime}$ = −0.32 V) with a suitable catalyst present, will there be any net conversion of pyruvate to lactate?

**Answer**

The concentration ratio of pyruvate/lactate is 1. Hence the redox potential of the

mixture will be the same as the standard redox potential (logarithm of 1 in Equation (4.8) is zero). The redox potential of the $NAD_{ox}/NAD_{red}$ mixture (a two-electron reaction) will be very close to −0.29 V (Note from Table 4.2 , that a ratio of oxidised to reduced of 90/10, will make the mixture approximately 30 mV more positive than the standard potential). The difference in redox potential will therefore be 0.1 V. The free energy change for the conversion will be negative and pyruvate will be converted to lactate provided any activation barriers can be overcome by a suitable catalyst.

value. The equation shows this is the case. When the ratio of oxidised/reduced has a value greater than 1, the logarithm of the ratio will be positive (the logarithm of a number greater than 1 is a positive number). Hence $E$ will be more positive than $E°$, the expected result. Conversely, when the reductant is at a larger concentration than the oxidant, the ratio will be less than 1 and the logarithm will be negative (the logarithm of a number less than 1 is a negative number). $E$ will then be more negative than $E°$.

**Table 4.1**  Effect of changing the ratio of oxidant to reductant on the redox potential of a mixture of the two forms

| Change in ratio $\dfrac{\text{oxidised}}{\text{reduced}}$ | Change in redox potential for one-electron reaction (mV) | Change in redox potential for two-electron reaction (mV) |
| --- | --- | --- |
| ×10000 | +180 | +90 |
| ×1000 | +120 | +60 |
| ×10 | +60 | +30 |
| ×1 | no change | no change |
| ×0.1 | −60 | −30 |
| ×0.01 | −120 | −60 |
| ×0.001 | −180 | −90 |

### 4.1.1  *Standard redox potentials*

How are the sign and value of $E$ determined? This is done by reference to the hydrogen redox couple which provides a standard against which all other couples can be related. The redox scale is defined relative to the hydrogen electrode in which hydrogen gas is bubbled over finely divided platinum immersed in a solution of acid at 1 M concentration (Figure 4.2). The half reaction for the hydrogen couple is

$$H_2 = 2H^+ + 2e^- \qquad (E^\circ = 0\,V)$$

Other redox couples are characterised by their tendency to donate or accept electrons from the hydrogen electrode. Those which accept electrons will be at a more positive potential and those which donate electrons will be at a more negative potential (Table 4.2). The voltage which has to be applied to prevent any electron transfer taking place, under standard conditions, provides a measure of the standard redox potential $E^\circ$ of the redox couple. Note that the hydrogen electrode involves $H^+$ as the oxidised form of hydrogen. If we want to express the redox potential of the hydrogen couple at pH 7 ($E^{\circ\prime}$), then from Equation (4.9),

$$E_{\text{hydrogen,pH7}} = 0 + \frac{8.314 \times 298 \times \ln(10^{-7})}{96\,485}$$
$$= -0.413\,V$$

## 4.2 Electron transfer reactions

**Inner sphere electron transfer** involves the transfer of electrons directly between redox centres or through a bridging ligand. An example is the direct reduction of molecular oxygen by copper and iron in cytochrome oxidase.

**Outer sphere electron transfer** involves transfer of electrons over longer distances (> 10 Å). The rate depends on distance, the molecular groups of the intervening medium, the relative geometry of the two centres and the redox potential difference between the centres.

Two main mechanisms have been defined to describe electron transfer reactions. *Inner-sphere electron transfer* takes place via one or more bridging ligands directly bonded between the two interacting centres. A bridged binuclear complex forms the pathway for direct electron transfer and distances are usually of the order of a few angstroms (1 Å = 0.1 nm). Such short-range transfer is rare in biology. It occurs in some oxidase enzymes where copper and iron are coordinated in the protein to form a binuclear centre for oxygen reduction. Much more common is *outer-sphere electron transfer*, which takes place between redox centres over greater distances which can be up to 30 Å. These redox reactions are very common in biology and are often coupled to other biological processes such as proton translocation across membranes. It is important in living systems that the electron transfer is reasonably fast and also specific, otherwise a short circuit would take place between the two redox centres and the free energy would be effectively lost. In most cases the actual pathway for the electron is not known. Electron transfer can occur in proteins through bonds with the protein itself acting as a large bridging ligand, or through spaces within the protein where the electron has to tunnel through the intervening medium from one group to the next. Experiments with haem proteins such as myoglobin and

*Figure 4.2*
**Redox potentials are defined relative to the standard hydrogen electrode**

In the hydrogen electrode, shown here in schematic form, a piece of platinum is suspended in a solution of 1 M $H^+$ ions and hydrogen gas is bubbled over it at 1 atmosphere partial pressure.

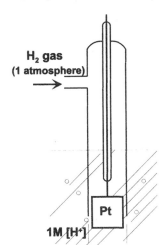

**H₂ gas
(1 atmosphere)**

**Pt**

**1M [H⁺]**

**Table 4.2**  Selected redox potentials for some biologically important reactions

| Redox couple | $n\,(e^-)$ | $m\,(H^+)$ | $E^{\circ\prime}$ (V) |
|---|---|---|---|
| hydroxylamine/ammonia | 2 | 0 | +0.90 |
| $O_2$ (g) /$H_2O$ | 4 | 4 | +0.82 |
| $Fe^{3+}/Fe^{2+}$ (inorganic) | 1 | 0 | +0.77 |
| nitrate/nitrite | 2 | 0 | +0.42 |
| nitrite/ammonia | 6 | 7 | +0.34 |
| $Fe^{3+}/Fe^{2+}$ (cytochrome *c*) | 1 | 0 | +0.25 |
| dehydroascorbate/ascorbate | 2 | 1 | +0.08 |
| nitrite/hydroxylamine | 4 | 3 | +0.07 |
| $N_2$ (g) /$NH_3$ | 6 | 6 | +0.06 |
| sulphate/sulphur ($S^0$) | 6 | 0 | +0.05 |
| CoQ/CoQH₂ | 2 | 2 | +0.04 |
| fumarate/succinate | 2 | 2 | −0.03 |
| oxaloacetate/malate | 2 | 2 | −0.17 |
| pyruvate/lactate | 2 | 2 | −0.19 |
| acetaldehyde/ethanol | 2 | 2 | −0.20 |
| FAD/FADH₂ | 2 | 2 | −0.22 |
| $S^0/H_2S$ | 2 | 2 | −0.23 |
| glutathione$_{ox}$/glutathione$_{red}$ | 2 | 2 | −0.23 |
| sulphur/hydrogen sulphide | 2 | 2 | −0.28 |
| $NAD(P)^+/NAD(P)H$ | 2 | 1 | −0.32 |
| acetoacetate/$\beta$-hydroxybutyrate | 2 | 2 | −0.35 |
| $2H^+/H_2$ | 2 | 2 | −0.41 |
| ferredoxin$_{ox}$/ferredoxin$_{red}$ | 1 | 0 | −0.43 |
| sulphate/sulphite | 2 | 1 | −0.52 |
| 3-phosphoglycerate/glyceraldehyde 3-phosphate | 2 | 2 | −0.55 |
| acetate/acetaldehyde | 2 | 2 | −0.60 |
| acetate/pyruvate | 2 | 2 | −0.70 |

cytochrome $c$, using photosensitive electron donors attached to the surface, have shown that electron transfer from the protein surface to the haem-iron takes place at a rate of approximately $30\,s^{-1}$ over a distance of around 12 Å. In contrast, the rate between cytochrome $c$ and its natural electron acceptor cytochrome-$c$ peroxidase is around $10^4\,s^{-1}$ over the longer distance of 17 Å. It is clear that factors such as distance, redox potential and the nature of the intervening medium have a strong influence on the reaction rate.

### 4.2.1 *Marcus theory of electron transfer*

A theoretical treatment of electron transfer by Marcus (1992 Nobel Laureate) makes predictions about the dependence of electron transfer rates on the properties of the redox systems involved. The rate ($k_{et}$) is predicted to fall exponentially with distance ($d$) and to depend on the nature of the intervening medium:

$$k_{et} = k_0 \exp(-\beta d) \qquad (4.11)$$

The value of the coefficient $\beta$ depends on the medium between the donor and acceptor and has the units of reciprocal distance. The rate falls most when the electron has to tunnel through a vacuum, with a value of $\beta$ of 2.8 Å$^{-1}$. It is easier for the electron to transfer when there are intervening $\pi$ and $\sigma$ orbitals between the two redox centres. In this case $\beta$ is around 1.2–1.6 Å$^{-1}$. At van der Waals contact, $\beta$ approaches 0.7 Å$^{-1}$ and the maximum electron transfer rate is around $10^{13}\,s^{-1}$ .

The second main factor influencing the electron transfer rate is the driving force for the reaction ($\Delta G$). However, the dependence of the rate on $\Delta G$ is not a simple one and shows the importance of the molecular nature of the reacting groups and their immediate environment. In simple reactions of the type formalised by Equation (4.1), no bonds are broken or formed. However there are usually differences between the oxidised and reduced forms of a redox centre in terms of ionic size, the orientation of neighbouring dipole moments and the arrangement and number of associated solvent molecules. There will be a particular arrangement which represents a minimum of the Gibbs free energy. Fluctuations around this minimum can be represented by a plot of the Gibbs free energy against the 'reaction coordinate' as shown in Figure 4.3. For electron transfer between two redox couples we have to consider both reactants and products. Since electrons move much faster than the rates of nuclear reorganisation, there is no time for nuclear motion such as dipole realign-

*Figure 4.3*

**Energy plots for electron transfer between two redox centres**

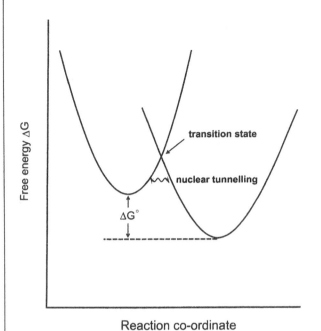

where the Gibbs free energies of the donor and acceptor are equal (the transition state). Quantum mechanics also predicts that electron transfer can take place by 'tunnelling', where the energy barrier can be crossed at a lower point along the curves where the nuclear wave functions overlap. However, in both cases, the greater the difference in nuclear and solvent arrangements between the centres, the more one centre will be displaced from the other (along the abscissa) and the greater will be the activation energy required for electron transfer. In contrast, the greater the difference in Gibbs energy between the centres (displacement along the ordinate) the less will be the activation energy required for electron transfer. Hence there will be a relationship between activation energy, which determines rate, and the thermodynamic parameter Gibbs free energy.

The Gibbs free energy of each centre depends on the position and configuration of the centre in its environment. Electron transfer will only take place at a configuration

ment during the electron transfer reaction and therefore electron transfer has to take place when the reactant and product have the same energy state (the *Franck–Condon principle*). This is where the two energy curves cross, the intersection point representing the reaction transition state. Marcus has derived an expression for the dependence of the rate constant for electron transfer ($k_{et}$) on the difference in free energy between the reactants and products ($\Delta G°$), and the amount of nuclear reorganisation that has to take place (the *reorganisational energy* $\lambda$) in order to match the energy states of the reactants and products:

$$k_{et} = A \exp\left[\frac{-\lambda}{4RT}\left(1 + \frac{\Delta G°}{\lambda}\right)^2\right] \qquad (4.12)$$

where $A$ is a factor which depends on the nature of the quantum mechanical coupling between the donor, acceptor, and molecules in the intervening environment. Taking the logarithm of both sides we have

$$\ln k_{et} = \ln A - \frac{1}{4\lambda RT}(\lambda + \Delta G^\circ)^2 \qquad (4.13)$$

Equation (4.13) has some interesting features. It predicts that the rate will be maximum when $\Delta G^\circ = -\lambda$, that is when the ground state of the reactant is the same as the transition state (curve b in Figure 4.4). The rate will then be temperature independent and will depend only on the electronic factor $A$. The reaction will slow down as $\Delta G^\circ$ becomes more positive (curve a in Figure 4.4) but will also slow down as $\Delta G^\circ$ becomes more negative (curve c in Figure 4.4). This last prediction seems to run counter to common sense. It states that the electron transfer rate will slow down as the driving force for the reaction becomes bigger. Nevertheless, experimental measurements have verified the prediction. In fact, the Marcus 'inverted' region may have some practical function in the primary reactions of photosynthesis, inhibiting the

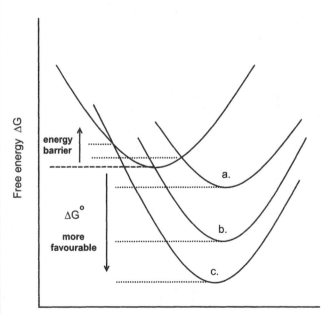

*Figure 4.4*

**Dependence of electron transfer on the driving force ($\Delta G$) and the position of the transition state**

The rate of electron transfer will depend on the height of the energy barrier between the reacting centres. The rate will be maximum when the transition state is at a position equal to the minimum of Gibbs free energy for the donor (curve b) and decreases as the transition state rises either side of the free energy minimum (curves a and c). This is despite the fact that the driving force for the reaction ($\Delta G$) becomes more favourable.

Free energy $\Delta G$

energy barrier

$\Delta G^\circ$

more favourable

a.

b.

c.

Reaction co-ordinate

back reaction of electron transfer following light-excitation of an electron from the bacteriochlorophylls (see Chapter 8).

## 4.3 Electron transfer complexes

In the cell, many electron transfer reactions are catalysed by large protein complexes containing a variety of redox cofactors. Electron flow from one complex to another is often linked by smaller proteins such as ferrodoxin and cytochrome $c$. Quinol compounds, with their hydrocarbon-like side-chains, catalyse electron transfer in the apolar environment of the membrane. The transport of electrons along a sequence of electron transfer complexes is not simply determined by redox potential. There has to be *specificity of interaction* between the redox components. Without this specificity, electron transfer would take place immediately to the final oxidant and all the free energy would be released in one reaction. In fact, *electron transport chains* catalyse electron flow in small steps between individual components with a controlled release of free energy from separate redox reactions. The main mechanism for coupling the redox energy to energy-requiring reactions is by the formation of proton gradients across the membrane of the electron transport chain (see Chapter 7).

### 4.3.1  *Mitochondrial electron transport chains*

Electrons from $NAD_{red}$, succinate and fatty acyl-CoA are transferred to molecular oxygen via a sequence of protein complexes embedded in or associated with the mitochondrial inner membrane (Figures 4.5 and 4.6). The specificity of interaction is such that the sequence of redox components closely follows the sequence of standard redox potential values. Thus electron flow from $NAD_{red}$ $(E^{\circ\prime} = -0.32\,V)$ to oxygen $(E^{\circ\prime} = +0.82)$ occurs via flavoprotein $(E^{\circ\prime} \approx -0.30\,V)$, iron–sulphur protein complexes $(E^{\circ\prime} \approx -0.4\,V$ to $0\,V)$, ubiquinone $(E^{\circ\prime} \approx 0\,V)$, $b$-type cyto-chromes $(E^{\circ\prime} \approx -0.05\,V)$, $c$-type cytochromes $(E^{\circ\prime} \approx +0.25\,V)$, $a$-type cytochromes $(E^{\circ\prime} \approx +0.3\,V)$ and copper $(E^{\circ\prime} \approx +0.3\,V)$. A short circuit between flavin and oxygen is avoided mainly by specific interactions between the redox components. Some direct 'leakage' to oxygen does take place, mainly from complex III. Oxygen intermediates such as superoxide and hydrogen peroxide can escape and could damage cellular structures if the protective enzymes *superoxide dismutase* and *catalase* were not present. Three of the compexes are intimately involved in energy coupling. These are com-

*Figure 4.5*
**Electron transfer complexes of mammalian
mitochondria**

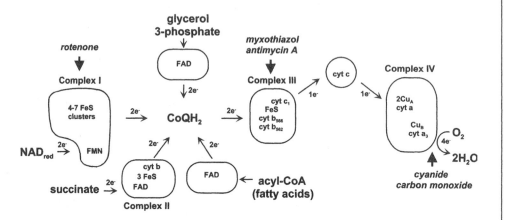

The electron transfer chain is a collection of electron transfer complexes which associate with or are embedded in the inner mitochondrial membrane. The lipid-soluble molecule coenzyme Q (CoQ) transfers electrons between the dehydrogenases and complex III. Water-soluble cytochrome *c* shuttles electrons between complexes II and IV. The pathway of electron transfer is determined by the redox potential of the complexes and the specificity of interaction between the components. All the dehydrogenases with the exception of glycerol-3-phosphate dehydrogenase have their substrate-binding site facing the matrix side of the inner membrane. (Thick arrows indicate the site of inhibitor action.)

pexes I, III and IV. Details of the mechanism of energy coupling are covered in Chapter 7.

### 4.3.1.1  *Complex I. NAD$_{red}$/coenzyme Q reductase*

This is the largest of the electron transport chain complexes, having up to 41 subunits with a total molecular mass of around 900 kDa. It catalyses the transfer of electrons from NAD$_{red}$ to ubiquinone and at the same time couples the redox energy to the transfer of protons across the mitochondrial inner membrane. It comprises two distinct complexes:

- A peripheral membrane complex, coded for by the nuclear genome, containing the NAD binding site, FMN and at least three iron–sulphur centres.

- An intrinsic membrane complex, coded for by the mitochondrial genome, with at least one iron–sulphur cluster and the ubiquinone binding site.

Although closely interacting in the final structure, both complexes are assembled separately and appear to have evolved

*Figure 4.6*
**Freeze-fracture electron micrograph of mitochondria**

In cross section (left) the outer and inner mito-chondrial membranes can be seen. The fold-ings of the inner membrane are known as cristae. The outer membrane of the mitochon-drion on the right has fractured to show a region of inner membrane in which are embedded the electron transport complexes. (From Wrigglesworth, J.M., Packer, L. and Branton, D., 1970, *Biochim. Biophys. Acta* **205**, 125–135.)

independently. In fact, sequence comparisons have shown the peripheral complex to be related to bacterial NAD-linked hydrogenases. The isolated peripheral complex has been shown to have some $NAD_{red}$–ubiquinone activity which is insensitive to piericidin, an inhibitor of the intact complex I. The suggestion has therefore been made that the peripheral and membrane structures may be linked by an internal ubi-quinone pathway. The detailed mechanism of coupling elec-tron transfer to proton translocation in complex I is unclear, but a ubiquinone/ubiquinol cycle similar to that catalysed by complex III has been put forward as a working hypothesis.

### 4.3.1.2   *Complex III. The cytochrome* $bc_1$ *complex*

The cytochrome $>bc_1$ complex catalyses electron transfer from reduced ubiquinone (ubiquinol) to cytochrome $c$ in the mitochondria of animals and plants. In plants, cyto-chrome $c_1$ is replaced by cytochrome $f$ (a $c$-type cytochrome called $f$ for historical reasons), giving the name $bf$ to the plant complex. The multisubunit complex contains three different redox proteins. A cytochrome $b$, comprising a single polypeptide with two $b$-type haems of different redox properties, a tightly bound $c$-type cytochrome termed $c_1$, and a 2Fe–2S iron–sulphur cluster (the Rieske protein). Cytochrome $b$ spans the membrane with eight $\alpha$-helical segments (four in the $bf$ complex) and a further helical segment is arranged along the cytoplasmic side of the inner membrane. The other subunits are attached around the central core of the cytochrome $b$. The linking of electron flow to proton translocation in complex III is the best understood of all the proton translocation mechanisms of the respiratory chain and is described in Chapter 7.

### 4.3.1.3  Complex IV. Cytochrome oxidase

Cytochrome oxidase is the terminal electron transfer complex of all eukaryotes and some prokaryotes where oxygen is used as a terminal acceptor. The crystal structure is known for the beef heart enzyme and for the enzyme from *P. denitrificans*. The beef heart enzyme has 13 subunits but fewer are found in other organisms. Three of the subunits, termed I, II and III, are coded for by the mitochondrial genome and these three are also present in prokaryotes. The remaining subunits are nuclear-coded, presenting the eukaryotic cell with an organisational problem in assembling the final structure. The function of the nuclear-coded subunits is not known, but almost certainly some are involved in regulation. Subunit I contains three of the four redox-active metal centres, one copper atom ($Cu_B$) and two haems ($a$ and $a_3$). These accept electrons from subunit II, which in turn accepts electrons one at a time from cytochrome $c$. The redox centre in subunit II contains two copper atoms in close interaction. This unusual centre stores only one electron at a time and for many years the presence of the second copper atom remained undetected. Overall, the fully reduced complex stores four electrons which are required for the full reduction of molecular oxygen to water. In fact the catalytic cycle does not involve the fully reduced enzyme but does involve an initial two-electron transfer to molecular oxygen to avoid the energy problem of superoxide formation (see Figure 6.5). It can be appreciated that different stages in the oxygen reduction reaction by the complex require different electron stoichiometries. Thus, electrons enter the complex one at a time from the single-electron donor cytochrome $c$. They are passed two at a time to a *binuclear centre* formed by haem $a_3$ and $Cu_B$ in close association. This site serves as the oxygen reduction centre. Release of one molecule of water leaves one oxygen atom tightly bound to haem $a_3$. At this stage, tight binding is essential if the oxygen atom is not to escape and react with other cellular components. Further two-electron transfer releases the oxygen as water. The overall reaction also requires four protons:

$$4\text{cyt}\,c^{2+} + 4\text{H}^+ + O_2 \rightarrow 4\text{cyt}\,c^{3+} + 2H_2O$$

Some of the free energy of the reaction (with a redox gap from cytochrome $c$ to oxygen of over 0.4 V) is conserved by the formation of a proton gradient across the inner mitochondrial membrane. A component of this gradient is formed by the consumption of protons in the oxygen reaction. These are taken from the mitochondrial matrix compartment. Further proton translocation is catalysed by a proton pump of as yet unknown mechanism. Further details of the energetics of the cytochrome oxidase reaction are dealt with in Chapter 7.

### 4.3.1.4  *Other electron transfer complexes of the respiratory chain*

Electrons from succinate are directly accepted by the mitochondrial electron transport chain via *succinate dehydrogenase* (termed complex II). This membrane-associated complex transfers two electrons from succinate (succinate/fumarate $E^{\circ\prime} = -0.03$ V) to ubiquinone ($E^{\circ\prime} \approx +0.04$ V) and is not involved in proton translocation. The redox cofactors in the complex are flavin as FAD, three iron–sulphur centres and a $b$-type cytochrome. One of the iron–sulphur centres has a standard redox potential much more negative than the flavin, which has provided a puzzle as to whether it is involved in electron transfer at all. A low-potential centre would be difficult to reduce by a more positive succinate or flavin couples. However, it should be remembered that the standard potential is no real guide to the function of an electron transfer component. The actual potential (determined by Equation (4.9)) could very well be more positive. Of course, from Equation (4.9), this would then mean that the steady-state reduction level would have to be very low ($< 0.01\%$) to bring the potential up to around 0 V.

Fatty-acyl-CoA also donates electron pairs to a membrane-bound dehydrogenase during the $\beta$-oxidation of fatty acids in mitochondria (see Chapter 6). The *acyl-CoA dehydrogenase* contains two flavins (as FAD) as well as an iron–sulphur centre. The electrons are finally transferred to ubiquinone. As with succinate dehydrogenase, this complex associates with the matrix side of the mitochondrial inner membrane and is not involved in proton translocation.

Mitochondria also have a membrane-bound *glycerol-3-phosphate dehydrogenase* on the cytoplasmic side of the inner membrane. This serves an important function in transfering reducing equivalents from the cytoplasm to the mitochondrial electron chain, the *glycerol phosphate shuttle* (Figure 4.7). $NAD_{red}$ generated in glycolysis (see Chapter 6) can be reoxidised by a cytoplasmic glycerol-phosphate dehydrogenase:

$$\text{dihydroxyacetone phosphate} + NAD_{red} \rightarrow \text{glycerol phosphate} + NAD_{ox} \tag{4.14}$$

The glycerol phosphate is then reoxidised back to dihydroxyacetone phosphate by the mitochondrial enzyme which donates the electrons into the electron transport chain rather than back to $NAD_{ox}$.

### 4.3.2 *Electron transport chains in plants*

Plant mitochondria contain additional redox complexes in their electron transport chains (Figure 4.8). In particular, most plant mitochondria contain a cyanide-insensitive

---

*Figure 4.7*
**The glycerophosphate shuttle**

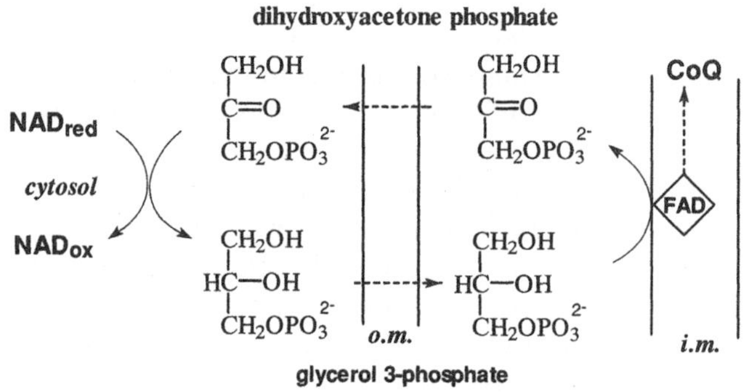

NAD$_{red}$ generated by redox reactions in the cytoplasm can donate electrons to the electron transport chain in the mitochondrial inner membrane (*i.m.*) using the glycerol-3-phosphate shuttle. A cytoplasmic glycerol-3-phosphate dehydrogenase catalyses the formation of glycerol 3-phosphate from dihydroxyacetone phosphate (an intermediate in glycolysis). The glycerol-3-phosphate is reconverted back to dihydroxyacetone phosphate by the membrane-associated dehydrogenase. The net result is the transfer of electrons from cytoplasmic NAD$_{red}$ to the mitochondrial electron transport chain. Unfortunately, electron entry into the chain is at the level of coenzyme Q and the free energy available is less than that from the oxidation of mitochondrial NAD$_{red}$.

*Figure 4.8*
**The electron transport chain of plant mitochondria**

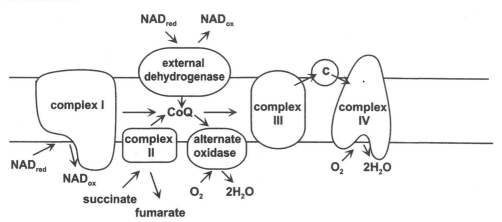

In addition to the main complexes found in animal mitochondria, plants often possess additional dehydrogenases both external and internal to the mitochondrial inner membrane. An alternative non-haem oxidase can oxidise ubiquinol releasing the redox energy as heat.

terminal oxidase in addition to the haem/copper-containing cytochrome oxidase. The *alternative oxidase* contains non-haem iron and is present at different concentrations in different plants. It is insensitive to cyanide but can be inhibited by hydroxamic acids. The free energy from the redox reaction is released as heat. Electron flux through the alternate oxidase pathway is particularly active in the spadices of *Arum* lilies and skunk cabbage, where the temperature of the spadex is raised by several degrees to aid the volatilisation of insect attractants (Figure 4.9).

*Figure 4.9*
**Heat generation by the *Arum* lily**

The temperature of the spadex of the *Arum* lily can be raised 10–15°C by oxidation of mitochondrial substrates through the alternate oxidase pathway. This aids the volatilisation of aromatic amines for insect attraction.

Plant mitochondria also possess external $NAD_{red}$ dehydrogenases. These are insensitive to rotenone, unlike the complex I dehydrogenase, and bypass the first free-energy-coupling site of the respiratory chain.

### 4.3.3 *Bacterial respiratory chains*

Variation in bacterial respiratory chains reflects the variation in habitat. As described in Chapter 5, some bacteria can survive on fermentation without the use of any respiratory chain at all. Others use respiratory chains with inorganic compounds such as sulphate and nitrate as terminal electron acceptors in place of oxygen (see Chapter 6). Many can switch between different electron donors and acceptors according to the growth conditions (Figure 4.10). One problem always faced, however, is that free energy conservation drops as the redox gap becomes less. For example, the growth of facultative bacteria under anaerobic conditions is slower with less *growth yield* (grams dry weight of organism per mole of substrate) than growth under aerobic conditions. Having oxygen as the final electron acceptor always maximises the available free energy.

> Bacterial electron transfer chains adapt to environmental conditions.

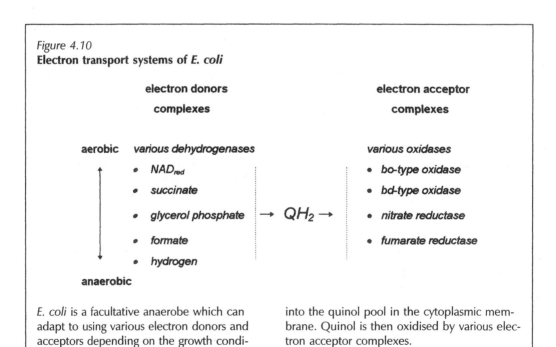

Figure 4.10
**Electron transport systems of *E. coli***

*E. coli* is a facultative anaerobe which can adapt to using various electron donors and acceptors depending on the growth conditions. Various dehydrogenases feed electrons into the quinol pool in the cytoplasmic membrane. Quinol is then oxidised by various electron acceptor complexes.

Most terminal oxidases of bacteria use ubiquinol as substrate rather than cytochrome *c* although cytochrome *c* oxidases are found in some species such as *Paracoccus denitrificans*. Many of the quinol oxidases have sequence homology to the cytochrome *c* oxidases. Common features include the presence of a copper/haem binuclear centre which acts as the site for oxygen reduction, and the ability of both type of oxidases to couple electron flow to proton translocation. There are differences in the electron donation site, not the least being the fact that cytochrome *c* is a one-electron donor whereas ubiquinol is a two-electron donor. A further ubiquinol oxidase, the *bd*-type oxidase, is also found in many bacteria. It contains three haems, two *b*-type and one *d*-type, but no copper. This type of oxidase has high affinity and activity for oxygen. In nitrogen-fixing bacteria, it is thought to play a part in protecting the nitrogenase complex from the damaging effects of molecular oxygen.

## Summary

- Oxidation involves the removal of electrons from a compound and reduction involves the gain of electrons. The electron acceptor is known as the oxidant and the electron donor is known as the reductant. The oxidised and reduced forms of a compound form a redox couple. Redox reactions involve the transfer of electrons from one redox couple to another. Protons may or may not be involved.

- The tendency of a redox couple to accept or donate electrons from other redox couples is quantified by the redox potential of the couple. Redox potentials are measured relative to the hydrogen ($H^+/H_2$) couple. Redox potentials can be altered by changes in the concentration of the oxidised and reduced forms of the couple. For one-electron reactions, the redox potential changes by approximately 60 mV for every 10-fold change in the oxidised to reduced ratio

of the couple, becoming more positive if the oxidised form increases and more negative if the reduced form predominates. For two-electron reactions the potential changes by 30 mV.

- Electron transfer between redox centres can take place by direct transfer over very short (<1 nm) distances (inner-sphere electron transfer) or over longer distances through the intervening medium (outer-sphere electron transfer). The rate of transfer for outer-sphere processes depends on the distance between the redox centres, the nature of the intervening medium, and the driving force for the reaction (mainly determined by the redox potential difference but also including the energy required for the nuclear reorganisation of the donor and acceptor and their associated solvent molecules).

- Electron transfer reactions in biology are determined by the specificity of interaction between oxidant and reductant rather than by the standard redox potential difference of the two couples. A wide variety of electron carriers has evolved in biology. These include small molecules such as NAD and flavin as well as large redox-active complexes. Electron transfer in biology is closely controlled and generally takes place over small redox potential gaps. Many of the redox components are organised into electron transport chains to facilitate the sequential transfer of electrons in discrete stages. This arrangement is essential for coupling the free energy of the redox reactions to other reactions with unfavourable free energy changes.

## Selected reading

Bendall, D.S., ed., 1996, *Protein Electron Transfer*, Oxford: BIOS Scientific Publishers Ltd. (A collection of theory and experiments at an advanced level)

Brock, T.D., Smith, D.W. and Madigan, M.T., 1984, *Biology of Microorganisms*, New Jersey: Prentice-Hall. (Comprehensive description of microbial systems)

Farid, R.S., Moser, C.C. and Dutton, P.L., 1993, Electron transfer in proteins, *Curr. Opin. Struct. Biol.* **3**, 225–233. (Marcus theory applied to practical examples)

Keilin, D., 1970, *The History of Cell Respiration and Cytochrome*, London: Butler & Tanner. (Gives an authorative account of the history of cell respiration)

Kotz, J.C. and Purcell, K.F., 1991, *Chemistry and Chemical Reactivity*, Florida: Saunders College Publishing. (Contains a clear introduction to electrochemistry)

Tyler, D.D., 1991, *The Mitochondrion in Health and Disease*, New York: VCH Publishers. (Useful for reference and also contains a readable account of the history of mitochondrial research)

Weiss, H., Friedrich, T., Hofhaus, G. and Preis, D., 1991, The respiratory-chain NADH dehydrogenase (complex I) of mitochondria, *Eur. J. Biochem.* **197**, 563–576.

Wrigglesworth, J.M. and Baum, H., 1980, The biochemical functions of iron, Jacobs, A. and Worwood, M. (eds), *Iron in Biochemistry and Medicine*, vol. 2, pp. 29–86, London: Academic Press.

### Crystal structures

Iwata, S., Ostermeier, C., Ludwig, B. and Michel, H., 1995, Structure at 2.8 Å resolution of cytochrome *c* oxidase from *Paracoccus denitrificans*, *Nature* **376**, 660–669.

Tsukihara, T., Aoyama, H., Yamashita, E., Tomizaki, T., Yamaguchi, H., Shinzawa-Itoh, Nakashima, R., Yaono, R. and Yoshikawa, S., 1995, Structures of metal sites of oxidised bovine heart cytochrome *c* oxidase at 2.8 Å, *Science* **269**, 1069–1074.

## Study problems

1. The redox potential of the hydrogen electrode is generally given as 0 V in biochemistry books but as −0.413 V in others. Account for this difference. Write an equation for the pH dependence of the hydrogen electrode?

2. What is the redox potential of a 90% oxidised mixture of NAD?

3. Calculate the value of $\Delta G^{\circ\prime}$ for the oxidation of lactate by oxidised cytochrome *c*. (Use the standard redox values given in Table 4.2.)

4. In respiring mitochondria, cytochrome *c* ($E^{\circ\prime} = +0.25$ V) is around 10% reduced. What is its redox potential under these conditions? The addition of antimycin A lowers the steady-state reduction level to 1%. Calculate the new redox potential. What is the percentage change in flux rate of electrons through this region of the chain?

5. Cytochrome $a_3$ in cytochrome *c* oxidase has a standard redox potential of +0.37 V. What would be its steady-state reduction level during respiration when reacting with oxygen at a potential of +0.61 V?

6. An alternate oxidase of plant mitochondria accepts electrons from succinate at a potential of −0.05 V and reduces molecular oxygen to water (at a potential of +0.76 V). What would be the heat output from the oxidation of 1 mmol of succinate?

# 5  Redox Reactions in Metabolism

## 5.1 Oxidation states

Many exergonic reactions in metabolism involve redox reactions. Sometimes there is a change in the charge on a molecule, for example cytochrome $c^{2+}$ being oxidised to cytochrome $c^{3+}$, but often there is no obvious alteration in electrical charge, for example the oxidation of succinate to fumarate. Chemists have introduced the useful concept of *oxidation number* or *oxidation state* of individual atoms in a molecule. This is an artificial aid or 'bookkeeping' device to help keep track of redox reactions between, and even within, molecules and can be extremely useful when studying the bioenergetics of metabolism. The oxidation number of an atom is defined as the charge the atom would have if all its bonds were considered completely ionic (the bond orbitals being fully displaced towards one atom). The number is expressed relative to the free element, which is defined to have an oxidation number of zero. For example an atom of oxygen (the free element), has an oxidation number of 0. Saturating the outer valancy shell by the addition of two electrons to make $O^{2-}$ makes the ionic form of oxygen with an oxidation number of $-2$. Molecular oxygen, $O_2$, where two oxygen atoms share their valence electrons, has an oxidation number of $-4$. In other words, 4 electrons would have to be added to a molecule of oxygen to make each atom completely ionic.

$$\ddot{\underset{..}{O}} :: \ddot{\underset{.}{O}} \quad + \quad 4e^- \quad \longrightarrow \quad \ddot{\underset{..}{O}} : \quad + \quad : \ddot{\underset{..}{O}}$$

molecular oxygen         ionic oxygen<br>
(oxidation number $-4$)        (oxidation number $-2$)

Hydrogen usually forms compounds with other atoms that are more electronegative. It can be thought of as acting as a reductant. In that case, its oxidation number is $+1$:

$$H^{\bullet} \quad \longrightarrow \quad H^+ + e^-$$

hydrogen atom        hydrogen ion<br>
(oxidation number 0)        (oxidation number $+1$)

Oxygen in water has an oxidation number of $-2$, since we assign the pair of electrons in the covalent bond between oxygen and hydrogen to the more electronegative oxygen atom. The two hydrogen atoms will then each have oxidation numbers of $+1$:

$$\text{H} \quad :\overset{..}{\underset{..}{\text{O}}}: \quad \text{H}$$

$$(+1) \quad (-2) \quad (+1)$$

Remember that oxidation numbers are simply a formal aid and do not represent the actual bonding in the molecule, which is usually covalent rather than ionic. The advantage of using this formalism becomes apparent in metabolism when we examine the oxidation numbers of carbon. The carbon atom has four electrons in its outer shell:

$$\cdot \overset{.}{\underset{.}{\text{C}}} \cdot$$

In the formation of a carbon–carbon bond the electron pair is shared equally since each has the same electronegativity and thus no change in oxidation number takes place:

$$\cdot \overset{.}{\underset{.}{\text{C}}} : \overset{.}{\underset{.}{\text{C}}} \cdot$$

However, the addition of four hydrogens to the carbon (to form methane) changes its oxidation number to $-4$.

$$\text{H}$$
$$\text{H} \quad :\overset{..}{\underset{..}{\text{C}}}: \quad \text{H}$$
$$\text{H}$$

In contrast, addition of two oxygen atoms to carbon to form $CO_2$ will change the carbon oxidation number to $+4$.

$$(-2) \quad (+4) \quad (-2)$$

Table 5.1 shows some values for the oxidation numbers of carbon in various functional groups. We can now easily analyse metabolic reactions where the change of one functional group to another involves a change in the oxidation number of the carbon atoms. The reduction of a carbon–carbon double bond, for example in fatty acid synthesis, involves the addition of two electrons and two protons:

**Table 5.1**  Some values for the oxidation number of carbon in various functional groups

| Functional group ( — represents a single bond to another carbon atom) | | Carbon oxidation number |
|---|---|---|
| — CH$_3$ | methyl | −3 |
| H<br>\|<br>— C —<br>\|<br>H | saturated carbon–carbon | −2 |
| H<br>\|<br>— C = C —<br>\|<br>H | unsaturated carbon–carbon | −1 |
| OH<br>\|<br>— C —<br>\|<br>H | hydroxyl | 0 |
| — C $\diagup^H_{\diagdown O}$ | aldehyde | +1 |
| $\diagdown$C = O $\diagup$ | carbonyl | +2 |
| — C $\diagup^{OH}_{\diagdown O}$ | carboxyl | +3 |
| O = C = O | dioxide | +4 |

$$
\begin{array}{ccc}
\text{COO}^- & & \text{COO}^- \\
| & \xrightarrow[2\text{H}^+]{2e^-} & | \\
\text{CH} & & \text{CH}_2 \\
|| & & | \\
\text{CH} & & \text{CH}_2 \\
| & & |
\end{array}
\qquad (5.1)
$$

A **ketone** is characterised by the presence of a carbonyl group where the carbon atom is bonded to two other carbon atoms.

The two carbons change their oxidation number from $-1$ to $-2$. The standard redox potential for the reaction is around $0\,V$. A more negative redox potential (around $-0.3\,V$) is found for the reduction of a ketone to an alcohol. In this case the carbon changes its oxidation state from $+2$ to $0$.

$$
\begin{array}{ccc}
\mathrm{COO^-} & & \mathrm{COO^-} \\
| & 2e^- & | \\
\mathrm{C{=}O} & \xrightarrow{\phantom{2e^-}} & \mathrm{H{-}C{-}OH} \\
| & 2H^+ & |
\end{array}
\qquad (5.2)
$$

The reduction of a carboxylic acid to an aldehyde (a redox potential of around $-0.6\,V$) involves a change on oxidation number of the carbon from $+3$ to $+1$

$$
\begin{array}{ccc}
\mathrm{COO^-} & 2e^- & \mathrm{CHO} \\
| & \xrightarrow{\phantom{2e^-}} & | \qquad + H_2O \\
 & 2H^+ &
\end{array}
\qquad (5.3)
$$

## 5.2 $\beta$-Oxidation of fatty acids

$$
\begin{array}{c}
\mathrm{CH_3} \\
| \\
\mathrm{(CH_2)_n} \\
| \\
\longrightarrow \; \mathrm{C_{(\beta)}\,H_2} \\
| \\
\mathrm{C_{(\alpha)}\,H_2} \\
| \\
\mathrm{COOH}
\end{array}
$$

Oxidation of fatty acids involves an initial attack on the $\beta$-carbon (the third carbon from the carboxyl end of the chain.

Because of the highly reduced nature of the carbon in fatty acids, the oxidation of fat is often the main source of energy for many organisms. In humans, fat provides the largest energy store in the body and is the main fuel for muscles under normal conditions (see Chapter 2). Fatty acids are initially oxidised to acetyl-CoA by a process of $\beta$-oxidation, so called because the carbon chain is cleaved at the $\beta$-carbon to produce two-carbon acetyl fragments. In eukaryotes the process takes place in mitochondria and involves a stepwise oxidation of the $\beta$-carbon from an initial $-2$ oxidation state to a final $+2$ oxidation state (Figure 5.1). The end result of one turn of the cycle is a fatty acid shortened by two carbons. The two-carbon fragment is linked to the thiol group on the coenzyme A to form acetyl-CoA.

$$
\mathrm{CoA{-}SH} \; + \; \overset{\displaystyle O}{\overset{\displaystyle \|}{\sim C}}{-}CH_3 \; \longrightarrow \; \mathrm{CoA{-}S}{-}\overset{\displaystyle O}{\overset{\displaystyle \|}{C}}{-}CH_3
$$

**coenzyme A**        **acetyl group**                    **acetyl-CoA**

The alternative, to remove the two-carbon fragments in the form of acetic acid ($CH_3COOH$), would produce a molecule resistant to further oxidation. This is because an oxygen ester link is more stable to hydrolysis than the corresponding

## Figure 5.1
## β-oxidation of fatty acids

$$R-CH_2-\overset{\beta}{CH_2}-\overset{\alpha}{CH_2}-\overset{O}{\overset{\|}{C}}-SCoA$$

**Acyl CoA** (β-carbon oxidation state -2)

FAD$_{ox}$

FAD$_{red}$

**(i) Oxidation**

$$R-CH_2-\overset{H}{\underset{H}{C}}=\overset{O}{\overset{\|}{\underset{}{C}}}-SCoA$$

**Enoyl CoA** (β-carbon oxidation state -1)

$H_2O$

**(ii) Hydration**

$$R-CH_2-\overset{OH}{\underset{H}{\overset{|}{C}}}-CH_2-\overset{O}{\overset{\|}{C}}-SCoA$$

**Hydroxyacyl CoA** (β-carbon oxidation state 0)

NAD$_{ox}$

NAD$_{red}$

**(iii) Oxidation**

$$R-CH_2-\overset{O}{\overset{\|}{C}}-CH_2-\overset{O}{\overset{\|}{C}}-SCoA$$

**Ketoacyl CoA** (β-carbon oxidation state +2)

CoA-SH

**(iv) Thiolysis**

$$R-CH_2-\overset{O}{\overset{\|}{C}}-SCoA \qquad CH_2-\overset{O}{\overset{\|}{C}}-SCoA$$

**Acyl CoA** (shortened by two carbon atoms)      **Acetyl CoA**

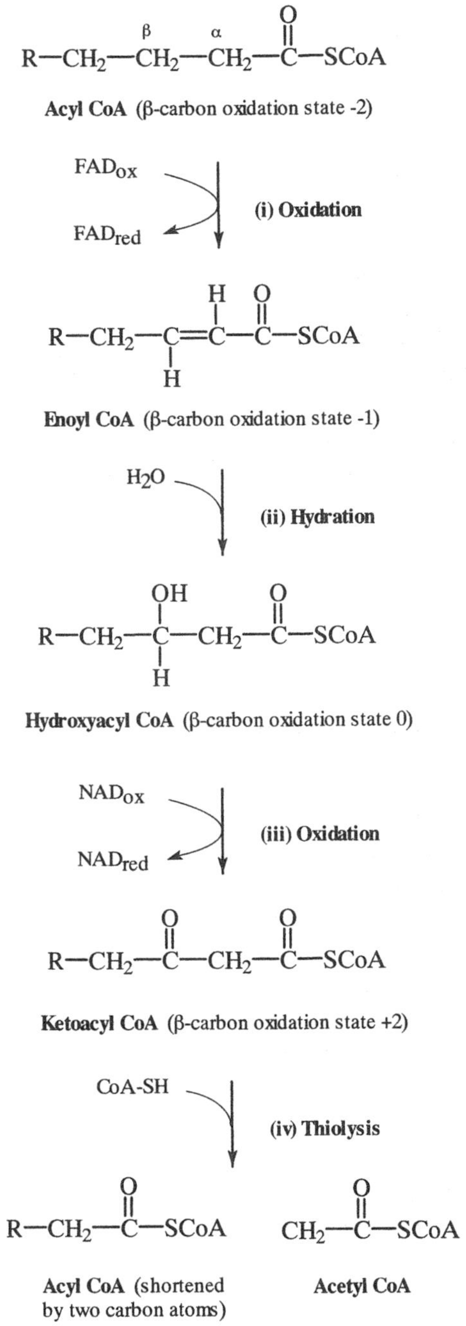

The oxidation of fatty acids takes place in a cyclic series of reactions each releasing a two-carbon fragment in the form of an acetyl group attached to CoA. The steps outlined for one cycle show that the β-carbon is successively oxidised from an initial −1 oxidation state to a final +2 state. The electrons from step (i) are accepted by the FAD prosthetic group of fatty-acyl-CoA dehydrogenase, whereas NAD accepts electrons from the oxidation step shown as (iii). Note that the oxygen added to the β-carbon in step (ii) comes from water.

$$R - C \overset{\overset{\displaystyle \delta^- O}{\|}}{\phantom{C}} \overset{\delta^+}{=} O^{\delta^+} - R'$$

**oxygen ester**

Resonance stabilisation results in a low $\Delta G°$ for hydrolysis.

$$R - C \overset{\overset{\displaystyle O}{\|}}{\phantom{C}} - S - R'$$

**thioester**

Lack of resonance stabilisation leads to high $\Delta G°$ for hydrolysis.

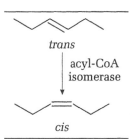

*trans*

acyl-CoA isomerase

*cis*

In fermentation, some substrate carbons are oxidised and others reduced. There are no external electron acceptors and therefore there is no net production of reducing equivalents.

thioester. The smaller atomic size of O compared with S increases the $\pi$-electron delocalisation around the carbon and contributes significantly to resonance stabilization. The $\Delta G°$ for hydrolysis is therefore low. Linking the acetyl group to coenzyme A by a *thioester bond* means that it can readily be transferred to other metabolites such as oxaloacetate in the first reaction of the citric acid cycle. Most fatty acids in nature have an even number of carbons and are oxidised to acetyl-CoA. However, those with an odd number give one molecule of propionyl-CoA and one of acetyl-CoA in the final cycle. Unsaturated fatty acids have to be altered before oxidation. These are usually in the *cis* configuration, which cannot be acted upon by the hydratase. They have first to be converted to the *trans* form by an isomerase before the cycle can continue. The electrons generated in $\beta$-oxidation are transferred to the mitochondrial respiratory chain by $NAD_{red}$ or directly by $FAD_{red}$ bound to fatty-acyl-CoA dehydrogenase. The main source of energy from fat comes from the subsequent oxidation of the acetyl group by the action of the citric acid cycle and the respiratory chain.

## 5.3 Oxidation of glucose by fermentation

*Glycolysis* (or the *Embden–Meyerhof glycolytic pathway*) is a sequence of reactions that convert the six-carbon molecule of glucose to two three-carbon molecules of pyruvate (Figure 5.2). Pyruvate can be metabolised in numerous ways depending on the conditions and the organism. In the absence of oxygen, it can be converted simply to lactate. This happens in mammals when the oxygen supply is limited as well as in many anaerobic microorganisms. The process is one of *fermentation*. Thus glycolysis is part of the fermentation of glucose to lactate. Fermentation is normally thought of as a series of catabolic reactions which provide an energy yield without any net oxidation or reduction. For example, the reducing equivalents produced by the oxidation of glucose to pyruvate, carried on $NAD_{red}$, are in turn used to reduce pyruvate to lactate. Thus a balance of oxidation and reduction is maintained. What does take place, however, is an internal redox change in the substrate as can be seen by a close examination of the carbon atoms in glucose and lactate:

*Figure 5.2*

**The glycolytic pathway**

The breakdown of glucose by the glycolytic pathway occurs in two main stages. The first, energy-requiring stage, involves a double phosphorylation of glucose followed by clea-vage to give two triose phosphates. The second stage involves the oxidation of an aldehyde to an acid coupled to the phosphorylation of ADP to ATP. Pyruvate is formed from phosphoenolpyruvate with a further phosphorylation of ADP.

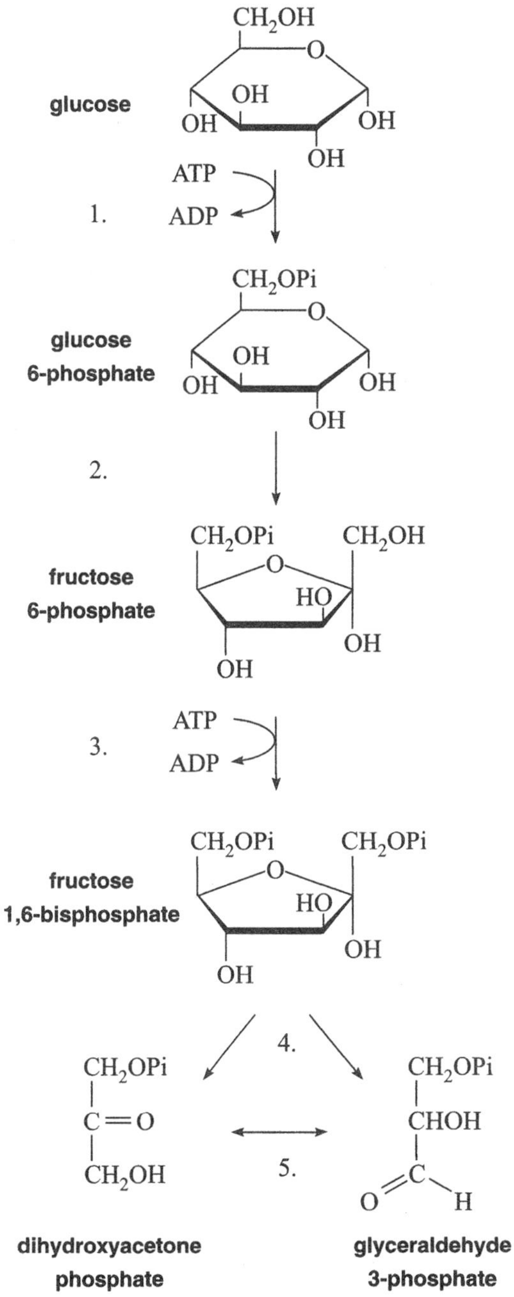

**Reaction**

1. The initial phosphorylation of glucose at the carbon-6 position by *hexokinase* traps the glucose within the cell and prepares the molecule for subsequent metabolism. An unfavourable phosphorylation reaction ($\Delta G^{\circ\prime} = +13.8\,\text{kJ}\,\text{mol}^{-1}$) is coupled to the favourable reaction of ATP hydrolysis ($\Delta G^{\circ\prime} = -31\,\text{kJ}\,\text{mol}^{-1}$) via an enzyme/intermediate complex (see Section 3.3). Overall $\Delta G^{\circ\prime} = -17.2\,\text{kJ}\,\text{mol}^{-1}$.

2. An interconversion catalysed by an *isomerase*. Readily reversible by slight changes in concentrations of reactants and products ($\Delta G^{\circ\prime} = +1.7\,\text{kJ}\,\text{mol}^{-1}$). The hydroxyl group on the carbon-1 position is now available for phosphorylation.

3. A second phosphorylation at the expense of the free energy of ATP hydrolysis is catalysed by *phosphofructokinase*, an enzyme with a primary role in the regualtion of glycolysis. Allosteric activators of the enzyme are AMP, ADP and the product fructose 1,6-biphosphate. ATP is an inhibitor. Overall, $\Delta G^{\circ\prime} = -14.2\,\text{kJ}\,\text{mol}^{-1}$.

4. Cleavage of the carbon-6 sugar into two triose phosphates is catalysed by *aldolase*. The breaking of the carbon–carbon bond is a very unfavourable reaction ($\Delta G^{\circ\prime} = +23.8\,\text{kJ}\,\text{mol}^{-1}$) and only takes place because the cellular concentrations of the products are kept low ($< \mu\text{m}$).

5. The normal equilibrium of the *isomerase* reaction favours dihydroxyacetone phosphate ($\Delta G^{\circ\prime} = +7.5\,\text{kJ}\,\text{mol}^{-1}$) but the product glyceraldehyde 3-phosphate is readily removed and pulls the reaction to the right. One glucose produces two triose phosphates and the stoichiometry of the following reactions of the pathway is therefore two per glucose molecule.

*(continued)*

*Figure 5.2*   **(continued)**

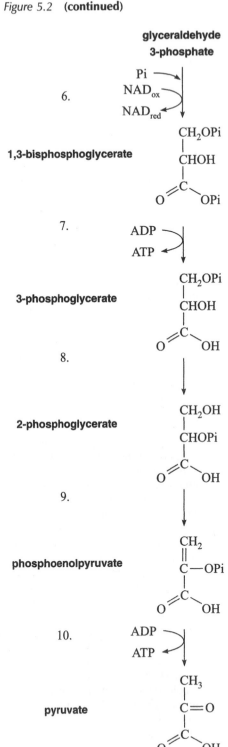

**6–7.** *Glyceraldehyde 3-phosphate dehydrogenase* catalyses the first free-energy yielding step of glycolysis. The energy of the oxidation of an aldehyde to an acid is coupled to the phosphorylation of ADP (see Figure 5.3). Overall $\Delta G^{\circ\prime} = -11.5\,\mathrm{kJ\,mol^{-1}}$.

**8.** Phosphate is transferred from the carbon-3, to the carbon-2 by *phosphoglycerate mutase* ready for a subsequent dehydration reaction to 'prime' the molecule for a phosphoryl transfer reaction. The reaction is close to equilibrium with an overall $\Delta G^{\circ\prime} = +4.6\,\mathrm{kJ\,mol^{-1}}$.

**9.** *Enolase* catalyses the dehydration of 2-phosphoglycerate to phosphoenolpyruvate. The latter has a highly unfavoured enol configuration which raises the phosphoryl transfer potential of the molecule to a high value, ready for phosphate transfer to ADP. Overall $\Delta G^{\circ\prime} = +1.7\,\mathrm{kJ\,mol^{-1}}$.

**10.** *Pyruvate kinase* catalyses the transfer of the phosphate on phosphoenolpyruvate to ADP with the formation of pyruvate. Pyruvate favours the keto form, shown here, rather than the enol configuration. Hence the reaction is pulled strongly to pyruvate formation with a free energy change sufficient to phosphorylate ADP. Overall $\Delta G^{\circ\prime} = -31.4\,\mathrm{kJ\,mol^{-1}}$.

**Glucose**  **2 × Lactate**

$$6 \times \left( \begin{array}{c} \text{OH} \\ | \\ -\text{C}- \\ | \\ \text{H} \end{array} \right) \longrightarrow 2 \times \left( \begin{array}{c} \text{H} \quad \text{OH} \\ | \quad | \quad\quad \text{O} \\ \text{H}-\text{C}-\text{C}-\text{C} \diagdown \\ | \quad | \quad\quad \text{OH} \\ \text{H} \quad \text{H} \end{array} \right)$$

| all carbons with oxidation state of 0 | reduced carbon, oxidation state altered to −3 | unaltered carbon | oxidised carbon, oxidation state altered to +3 |

'Internal' redox changes still yield energy. The standard free energy change for the fermentation of glucose to two molecules of lactate is $-217\,\text{kJ mol}^{-1}$, some of which is used for the formation of 2 moles of ATP per mole of glucose fermented. The coupling of ATP formation ($\Delta G^{\circ\prime} = +31\,\text{kJ mol}^{-1}$) to the breakdown of glucose (see below) means that the fermentation process has an overall standard free energy change of $-155\,\text{kJ mol}^{-1}$.

Net fermentation of glucose to lactate ($\Delta G^{\circ\prime} = -155\,\text{kJ mol}^{-1}$) glucose → 2 lactate 2ADP → 2ATP

## 5.4 Coupling redox reactions to ligand reactions

Redox reactions in the cell are often coupled to other reactions with an otherwise unfavourable free energy change. The mechanism of coupling varies, but a common example is where the redox reaction is linked to a ligand reaction by a chemical intermediate. The best-known example occurs in glycolysis, where the oxidation of glyceraldehyde 3-phosphate to 3-phosphoglycerate is coupled to to the formation of ATP. Often called *substrate-level phosphorylation*, the reaction could in fact be more accurately described as an oxidative phosphorylation since the favourable oxidation reaction of an aldehyde to an acid (the reverse of Equation (5.3)) is coupled to the unfavourable ligand reaction, the phosphorylation of ADP. The overall reaction can be broken down into three stages for thermodynamic analysis:

(i) The oxidation of glyceraldehyde 3-phosphate to 3-phosphoglycerate ($E^{\circ\prime} = -0.54\,\text{V}$)

(ii) The reduction of $\text{NAD}_{\text{ox}}$ to $\text{NAD}_{\text{red}}$ ($E^{\circ\prime} = -0.32\,\text{V}$)

(iii) The phosphorylation of ADP ($\Delta G^{\circ\prime} = +31\,\text{kJ mol}^{-1}$)

From Equation (4.4) ($\Delta G = -nF\,\Delta E$) we could calculate the energy release under standard conditions for the two half reactions in (i) and (ii). However, in the cell, the concentration of glyceraldehyde 3-phosphate is one hundred times less than that of 3-phosphoglycerate. The actual redox potential of the couple is therefore approximately 0.06 V more positive than the standard value (the standard redox potential will be changed by 0.03 V for every 10-fold change in the oxidised/reduced ratio, for a two-electron reaction, see Table 4.1). Similarly, the $NAD_{ox}/NAD_{red}$ ratio in the cytoplasm of the cell is very high, around $10^3$. The actual redox potential of the couple in (ii) will therefore be $-0.23$ V. Hence the overall free energy change in the oxidation of glyceraldehyde 3-phosphate will be

$$\Delta G' = -2 \times 96\,500 \times (0.48 - 0.23)$$
$$= -2 \times 96\,500 \times 0.25$$
$$= -48\,250\,\mathrm{J\,mol^{-1}}$$

The $ATP/[ADP][Pi]$ ratio in the cytoplasm is around 600. Using $\Delta G = \Delta G^\circ + RT \ln ([\text{products}] / [\text{reactants}])$ (Equation (3.3)), the energy required to drive the reaction in the direction of ATP formation will be

$$\Delta G' = +31\,000 + (8.314 \times 310 \times \ln 600)$$
$$= +47\,500\,\mathrm{J\,mol^{-1}}$$

Comparing the two values, it can be seen that there is just enough energy released in the redox reaction to drive ATP formation, although if the $NAD_{ox}/NAD_{red}$ ratio in the cytoplasm were to fall from its very oxidising value then glycolysis would be inhibited. (In the mitochondrion, the adenine nucleotide pool is at a much more reducing potential and this is presumably why the mitochondrial membrane is impermeable to $NAD_{ox}$ and $NAD_{red}$. Otherwise the cytoplasmic and mitochondrial pools would equilibrate and glycolysis would stop.) The common intermediate linking the redox reactions (i) and (ii) to the ligand reaction (iii) is the phosphorylated metabolite 1,3-bisphosphate dehydrogenase (Figure 5.3).

The cytoplasmic and mitochondrial $NAD_{ox}/NAD_{red}$ ratios are not in redox equilibrium.

Another phosphorylated metabolite is involved in the second ADP phosphorylating step in glycolysis. Phosphoenolpyruvate donates its phosphate group to ADP in a reaction catalysed by pyruvate kinase ($\Delta G^{\circ\prime} = -31\,\mathrm{kJ\,mol^{-1}}$). Unlike the glyceraldehyde 3-phosphate reaction, however, this is a straightforward ligand transfer reaction and does not involve redox changes. The reaction is favourable for ATP and pyruvate formation since the free energy change for phosphoenolpyruvate to pyruvate is very

*Figure 5.3*
**ADP phosphorylation in glycolysis**

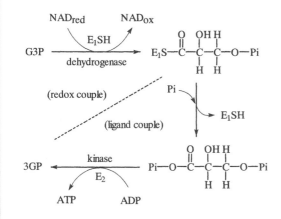

The oxidation of glyceraldehyde 3-phosphate to 3-phosphoglycerate is coupled to the phosphorylation of ADP via an intermediate metabolite 1,3-bisphosphoglycerate. The free energy change of the redox reaction is used to drive the ligand coupling reaction.

negative $(\Delta G^{\circ\prime} = -62\,\text{kJ}\,\text{mol}^{-1})$. The common intermediate in this case is a phosphorylated enzyme complex.

## 5.5 Fermentation pathways

Glycolysis is the most universal of metabolic pathways but it is not the only sequence for the breakdown of glucose to pyruvate. In most organisms, the *pentose phosphate pathway* is also present. Other names given to this sequence of reactions are the *hexose monophosphate shunt* and the *phosphogluconate pathway*. Its role is to produce pentose sugars for the biosynthesis of nucleic acids and is therefore most active in rapidly dividing cells. Its other role is to provide reducing equivalents in the form of $NADP_{red}$. The reductive stages of many biosynthetic reactions, for example fatty acid biosynthesis, are specific for $NADP_{red}$ rather than $NAD_{red}$. The two forms of the coenzyme, NAD and NADP, are not in redox equilibrium. Both have specific enzymes to catalyse their redox reactions. The cell is thus able to maintain different ratios of $NAD_{ox}/NAD_{red}$ to $NADP_{ox}/NADP_{red}$. In mammalian liver, the $NAD_{ox}/NAD_{red}$ ratio is around 700, whereas the $NADP_{ox}/NADP_{red}$ ratio is very much lower at 0.01.

The cytoplasmic ratios of $NAD_{ox}/NAD_{red}$ and $NADP_{ox}/NADP_{red}$ are not in redox equilibrium.

The sequence of reactions starts with the oxidation of glucose 6-phosphate (Figure 5.4). This has a very negative (favourable) free energy change and is catalysed by glucose 6-phosphate dehydrogenase, an enzyme highly specific for $NADP_{ox}$ rather than $NAD_{ox}$. It also is strongly inhibited by

*Figure 5.4*
**The pentose phosphate pathway for glucose metabolism**

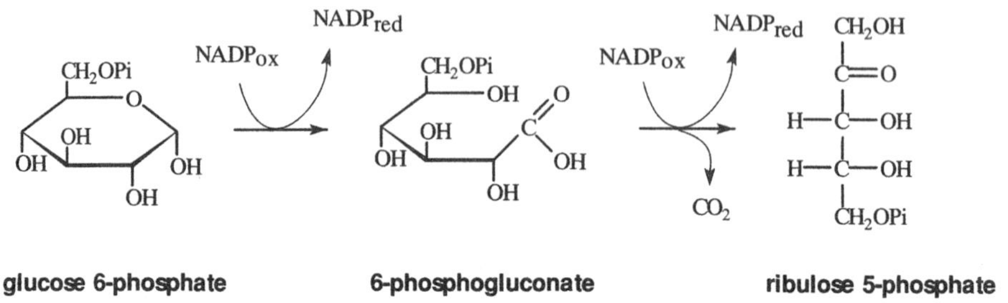

glucose 6-phosphate          6-phosphogluconate          ribulose 5-phosphate

The first sequence of reactions involve redox changes in the sugar carbons. The electrons are removed to produce two molecules of $NADP_{red}$ and one carbon is fully oxidised to $CO_2$. The ribulose 5-phosphate can then be metabolised to ribose 5-phosphate for nucleic acid biosynthesis or converted eventually into triose phosphates for further carbon oxidation in the later reaction stages of glycolysis.

$NADP_{red}$, which acts as the main controller of flux through the pathway. A further oxidation, also generating $NADP_{red}$, forms the five-carbon sugar ribulose 5-phosphate. In these reactions, one of the carbons on the original glucose is fully oxidised to $CO_2$. The ribulose 5-phosphate can then be converted to ribose 5-phosphate ready for nucleic acid biosynthesis. Overall, this stage of the pathway can be summarised as

$$\text{glucose 6-phosphate} + 2NADP_{ox} \rightarrow$$
$$\text{ribose 5-phosphate} + CO_2 + 2NADP_{red} \quad (5.5)$$

Alternatively, a series of reactions can convert the ribulose 5-phosphate into intermediates of glycolysis for further carbon oxidation.

The flux through the pentose phosphate pathway depends on the needs of the particular cell. Thus cells actively involved in biosynthesis have high activity and this is directed to ribose 5-phosphate production for nucleic acid biosynthesis. Some cells require $NADP_{red}$ for specific redox reactions, for example fatty acid biosynthesis. In this case the ribose 5-phosphate is converted into fructose 6-phosphate and glyceraldehyde 3-phosphate, both of which can proceed down the glycolytic pathway to pyruvate. The reaction sequence will then be:

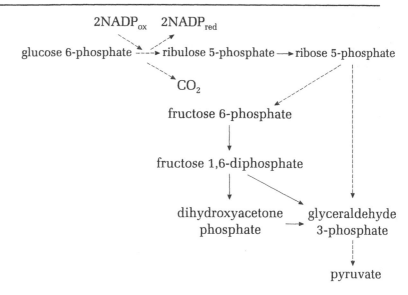

In bacteria, the percentage flux through the pentose phosphate pathway depends on the requirements of the organism and can change acording to growth conditions. For example, aerobic cultures of *E. coli* catabolise glucose mainly using the pentose phosphate pathway whereas, under anaerobic conditions, the glycolytic pathway predominates. Pentose sugars are often produced by the breakdown of many plant polymers and these can be used as fermentation substrates by various lactobacilli and enteric bacteria.

Many bacteria can also metabolise glucose via the *Entner–Doudoroff pathway* (Figure 5.5). The 6-phosphogluconate molecule produced in the first stage of the pentose phosphate pathway is rearranged by an internal redox reaction into 2-oxo-3-deoxy-6-phosphogluconate. This six-carbon molecule can then be split into two three-carbon molecules by aldolase to yield pyruvate and glyceraldehyde 3-phosphate. Because three of the carbon fragments are directly converted to pyruvate without phosphorylation, the pathway will only give 1 mole of ATP net per mole of glucose compared to the 2 moles of ATP in glycolysis. However, the pathway is important for bacteria metabolising gluconate and similar substrates (Figure 5.6).

Microorganisms can ferment a variety of other compounds. The substrate needs to be neither too oxidised nor too reduced, otherwise electron donation and acceptance will be inhibited. Hydrocarbons and fatty acids are nonfermentable. If another compound can be used as an electron acceptor, then mixed fermentation is possible. For example, in some species of clostridia, alanine is oxidised to acetate via a number of steps. At the same time glycine is reduced

*Figure 5.5*
**The Entner–Doudoroff pathway for glucose metabolism by bacteria**

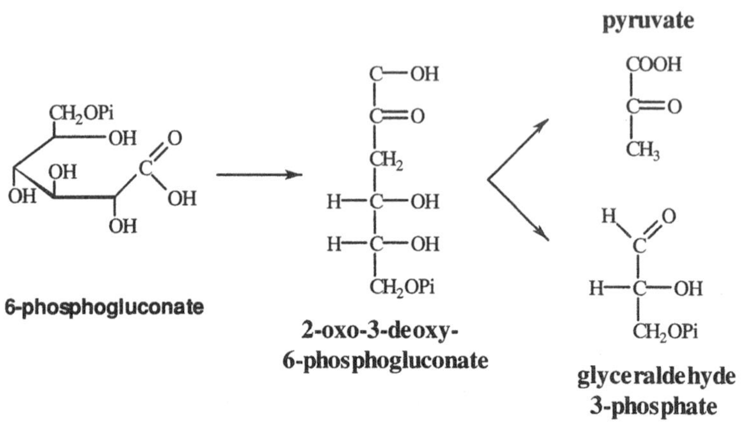

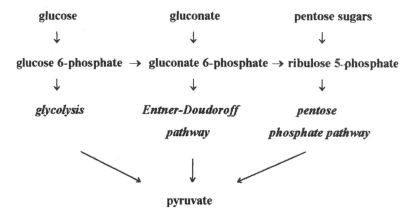

**6-phosphogluconate**

**2-oxo-3-deoxy-6-phosphogluconate**

**pyruvate**

**glyceraldehyde 3-phosphate**

Glucose and gluconate are often degraded by aerobic and anaerobic bacteria via the Entner–Doudoroff pathway. Because only one molecule of glyceraldehyde 3-phosphate is produced from the six-carbon sugar, the net yield of ATP by substrate-level phosphorylation is halved compared with glycolysis.

*Figure 5.6*
**Glucose and gluconate metabolism in bacteria.**

|    glucose    |    gluconate    | pentose sugars |
| :-----------: | :-------------: | :------------: |
|       ↓       |        ↓        |        ↓       |

glucose 6-phosphate  →  gluconate 6-phosphate → ribulose 5-phosphate

|       ↓       |        ↓        |        ↓       |
| :-----------: | :-------------: | :------------: |
|  *glycolysis* | *Entner-Doudoroff pathway* | *pentose phosphate pathway* |

**pyruvate**

The pathways utilised will depend on the substrates available and the energetic requirements of the organism.

to acetate. Such coupled redox reactions are known as *Stickland reactions* (see Table 5.2).

**Table 5.2** Some amino acids coupled in mixed fermentations

Fermentations by many bacteria can involve the oxidation of one substrate coupled to the reduction of a second substrate. These are known as mixed fermentations or Stickland reactions.

| Oxidisable substrate | Reducible substrate |
| --- | --- |
| alanine | glycine |
| leucine | proline |
| isoleucine | hydroxyproline |
| valine | tryptophan |
| histidine | arginine |

## 5.6 Metabolic fates of pyruvate in the absence of oxygen

Lactate is not the only end product of glucose fermentation, although it is the only one in mammals. In yeast, fermentation generates ethanol by the action of pyruvate decarboxylase (i) and alcohol dehydrogenase (ii) (Equation (5.6)).

$$
\underset{\textbf{pyruvate}}{CH_3-\overset{\overset{\textstyle O}{\|}}{C}-COO^-} \xrightarrow[\text{(i)}]{H^+\ CO_2} \underset{\textbf{acetaldehyde}}{CH_3-\overset{\overset{\textstyle O}{\|}}{C}-H}
$$

$$
\xrightarrow[\text{(ii)}]{NAD_{ox}\ NAD_{red}} \underset{\textbf{ethanol}}{CH_3-CH_2-OH}
$$

(5.6)

Note that one of the carbons of pyruvate ends up oxidised as $CO_2$ (oxidation state $+4$) while the others end up more reduced in ethanol (net oxidation state of $-4$). Again note that the reducing equivalents formed in pyruvate production are used in ethanol formation. The energy yield from fermentation reactions comes from these internal redox changes. It is clear that, in the absence of any external electron acceptor such as oxygen, the largest energy yield will occur when half the carbons are fully oxidised and end up as carbon dioxide ($CO_2$) and the other half end up fully reduced as methane ($CH_4$). Not all fermenting organisms are efficient enough to do this. For example, the anaerobic breakdown of pyruvate by clostridia produces $CO_2$ and acetate. In this case $H^+$ acts as an electron acceptor and molecular hydrogen is produced. Pyruvate is decarboxylated and acetyl-CoA accepts the resulting two carbon fragment to form acetyl-CoA.

$$\text{pyruvate} + \text{CoA} \rightarrow \text{acetyl-CoA} + CO_2 + H_2 \qquad (5.7)$$

The direct production of molecular hydrogen is an unusual reaction and requires the presence in clostridia of hydrogenase, an enzyme complex containing iron–sulpur clusters and nickel. The acetyl group is eventually released as acetate, but not before the free energy of the reaction is used to drive the phosphorylation of ADP to ATP. The coupling intermediate is the metabolite acetylphosphate ($\Delta G^{\circ\prime} = -42.3\,\text{kJ}\,\text{mol}^{-1}$).

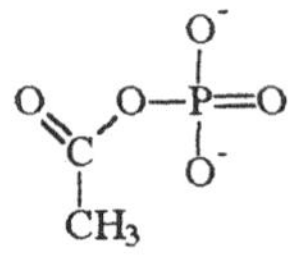

$$\text{acetyl-CoA} \quad + \quad P_i \longrightarrow \text{CoA} \quad + \quad \boxed{\text{acetylphosphate}} \quad (5.8)$$

$$\boxed{\text{acetylphosphate}} \quad + \quad \text{ADP} \longrightarrow \text{acetate} \quad + \quad \text{ATP} \qquad (5.9)$$

This sequence is known as the *phosphoroclastic reaction*. Other microorganisms in an anaerobic environment produce $CO_2$ and short-chain fatty acids such as butyrate and propionate. The gut of ruminants such as cows is full of fermenting bacteria and the fatty acid products from the fermentation of cellulose from grass are absorbed across the rumen wall to provide the main energy nutrition for the animal. One of the other products, methane, is expelled as flatulence and is thought to be a significant contributor to atmospheric methane levels and global warming by the greenhouse effect (see Box 8.1).

## 5.7 Metabolic fate of pyruvate in the presence of oxygen

In aerobes, the three carbons of pyruvate can be fully oxidised to $CO_2$, with the electrons ultimately reducing molecular oxygen. It should be noted that molecular oxygen does not provide the oxygen atoms for $CO_2$ formation. These come from water and the oxygen atoms originally in the sugar molecule. Oxygen acts only as an electron acceptor to form water as a final reduced product. The reaction sequence for full pyruvate oxidation can be broken down into three stages (Figure 5.7):

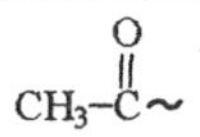

(i)   The oxidation of pyruvate ($C_3$) by pyruvate dehydrogenase to form an acetyl group ($C_2$) and one molecule of $CO_2$.

acetylphosphate

an acetyl group

*Figure 5.7*
**The three stages of pyruvate oxidation in
aerobic organisms**

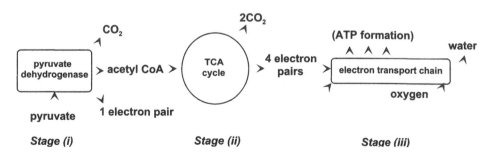

**Stage (i)**                    **Stage (ii)**                    **Stage (iii)**

The oxidation of pyruvate by aerobic metabolism involves (i) a decarboxylation in the pyruvate dehydrogenase reaction, (ii) oxidation of
the two remaining carbons to carbon dioxide
in the TCA cycle, and (iii) reduction of molecular oxygen by the reducing equivalents from
stages (i) and (ii). Free energy is conserved in
the electron transport chain reactions which
are coupled to ATP formation.

(ii)   The oxidation of the acetyl group to two molecules of
$CO_2$ in the citric acid cycle.

(iii)  The transfer of the electrons from (i) and (ii) to the
electron transport chain and then to molecular oxygen.

The reactions of the electron transport chain provide the
largest change in free energy and are coupled to ATP formation.

### 5.7.1   *Stage (i). Pyruvate dehydrogenase*

Pyruvate dehydrogenase is the link between glycolysis and
the citric acid cycle. In the overall reaction, pyruvate is decarboxylated to form acetyl-CoA. The electrons removed in
the redox reaction are used to form $NAD_{red}$. The multi-
enzyme complex has three activities, $E_1$, $E_2$ and $E_3$, each
requiring essential cofactors:

$E_1$   pyruvate dehydrogenase (decarboxylation of pyruvate, involving thiamine pyrophosphate);

$E_2$   dihydrolipoamide transacetylase (transfer of the
acetyl group to CoA involving lipoic acid);

$E_3$   dihydrolipoamide dehydrogenase (reduction of NAD
involving FAD).

The organisation of the three enzyme activities in a multi-
enzyme complex provides great efficiency by preventing

diffusion of the intermediates into bulk solution. The complex is allosterically inhibited by ATP, $NAD_{red}$ and acetyl-CoA but most importantly its activity is affected by phosphorylation and dephosphorylation. A specific kinase can phosphorylate a serine residue on $E_1$ to inhibit the complex. A corresponding phosphatase removes the bound phosphate to relieve the inhibition. The phosphatase is activated by $Ca^{2+}$ and $Mg^{2+}$, which in turn reflect the concentrations of free ATP. Thus when ATP is abundant and when the $NAD_{red}$ and acetyl-CoA levels are high, pyruvate dehydrogenase activity is turned off (Figure 5.8)

### 5.7.2 Stage (ii). The citric acid cycle

The *citric acid cycle*, or *tricarboxylic acid* (TCA) *cycle* as it is also known, is an efficient mechanism for oxidising the two carbons of the acetyl group attached to CoA. These two carbons are relatively stable to redox reactions when in the form of acetic acid and therefore they enter the cycle as an acetyl group on CoA. They are then easily attached to oxaloacetate, a four-carbon compound, to form citrate, a six-carbon compound. Citrate can then be converted back to oxaloacetate in a series of reactions, including two decarboxylations, with the release of two molecules of $CO_2$. The oxygen atoms are provided by water and the electrons from the redox reactions are collected by NAD or, in the case of succinate oxidation, passed directly to the electron transport chain. (It is somewhat analogous to the reactions in photosynthesis (Chapter 8), where water is split at the expense of light energy to provide reducing equivalents for metabolism.) The citric acid cycle only occurs under aerobic conditions since the

Figure 5.8
**Control of pyruvate dehydrogenase**

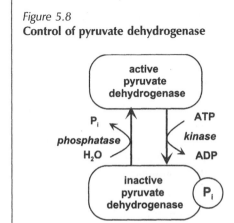

Pyruvate dehydrogenase is regulated by phosphorylation and dephosphorylation. A specific kinase phosphorylates the complex and inhibits its activity. Reactivation occurs when a specific phosphatase removes the bound phosphate. The two controlling enzymes are themselves regulated. When ATP/ADP or $NAD_{red}$/$NAD_{ox}$ ratios are high, and further free energy production is not required, the pyruvate dehydrogenase reaction is switched off.

activity of the electron transfer chain is necessary to regenerate oxidised NAD. In eukaryotes, the enzymes of the cycle are located in the mitochondrial matrix.

The individual reactions can be summarised with reference to Figure 5.9:

- *Citrate synthase* catalyses the formation of citrate from acetyl-CoA and oxaloacetate. Water is used for the release of free CoA. The reaction equilibrium is far in the direction of citrate production $(\Delta G^{\circ\prime} = -32.2\,\text{kJ}\,\text{mol}^{-1})$ because of the high energy of hydrolysis of acetyl-CoA $(\Delta G^{\circ\prime} = -36.8\,\text{kJ}\,\text{mol}^{-1})$ which makes this, the first step in the cycle, essentially irreversible.

- *Aconitase* catalyses the transfer of a hydroxyl group from C-3 to C-2 on citrate by successive dehydration and hydration reactions $(\Delta G^{\circ\prime} = +6.3\,\text{kJ}\,\text{mol}^{-1})$. Isocitrate, a secondary alcohol, is then more suitable for the subsequent oxidation reaction.

- *Isocitrate dehydrogenase* catalyses the NAD-linked oxidative decarboxylation of isocitrate to form 2-oxoglutarate $(\Delta G^{\circ\prime} = -20.9\,\text{kJ}\,\text{mol}^{-1})$.

- *2-Oxoglutarate dehydrogenase* catalyses the NAD-linked oxidative decarboxylation of $\alpha$-oxoglutarate $(\Delta G^{\circ\prime} = -33.5\,\text{kJ}\,\text{mol}^{-1})$. The reaction is very similar to that of pyruvate dehydrogenase and requires the same cofactors. However, instead of an acetyl group being transferred to CoA, a succinyl group is attached to form succinyl-CoA.

- *Succinyl-CoA synthase* catalyses the phosphorylation of GDP to form GTP. Looking at this and the previous reaction, it can be seen that we have another example of substrate-level phosphorylation where redox reactions are linked to ligand reactions. The formation of GTP is driven by the oxidation of 2-oxoglutarate to succinate with succinyl-CoA acting as the common intermediate linking the two reactions:

$$2\text{-oxoglutarate} + \text{NAD}_{\text{ox}} + \text{CoA} \rightarrow$$
$$\text{succinyl-CoA} + CO_2 + \text{NAD}_{\text{red}} \tag{5.10}$$

$$(\Delta G^{\circ\prime} = -33.5\,\text{kJ}\,\text{mol}^{-1})$$

$$\text{succinyl-CoA} + P_i + \text{GTP} \rightarrow$$
$$\text{succinate} + \text{GTP} + \text{CoA} \tag{5.11}$$

$$(\Delta G^{\circ\prime} = -2.9\,\text{kJ}\,\text{mol}^{-1})$$

The GTP can transfer its terminal phosphate group to ADP to form ATP by the action of *nucleoside diphospho-*

*Figure 5.9*
**The citric acid cycle**

In order to prepare for oxidative decarboxylation of citrate, aldolase catalyses the interchange of H and OH between positions 2 and 3

The first net oxidation reaction is accompanied by decarboxylation. Note that the carbon lost is not one of those present on the original acetyl group

A further oxidative decarboxylation in the presence of CoA produces a thiol-ester of succinate. Again, note that the carbon lost is not from the original acetyl group

Reactions of the citric acid cycle. The two carbons of the acetyl group on acetyl-CoA are attached to oxalo-acetate ($C_4$) to form citrate ($C_6$). Two carbons are lost by decarboxylation (in fact not the same carbons that entered on the acetyl group) and the resulting succinate ($C_4$) is oxidised in a series of reactions to regenerate oxaloacetate.

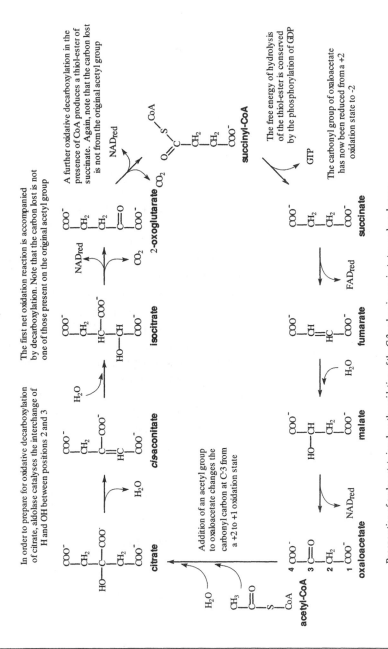

Addition of an acetyl group to oxaloacetate changes the carbonyl carbon at C-3 from a +2 to +1 oxidation state

The free energy of hydrolysis of the thiol-ester is conserved by the phosphorylation of GDP

The carbonyl group of oxaloacetate has now been reduced from a +2 oxidation state to -2

Regeneration of oxaloacetate involves the oxidation of the C-3 carbon in succinate to a carbonyl group *kinase.* Some bacteria can utilise ADP directly in the synthase reaction.

- *Succinate dehydrogenase* catalyses the FAD-linked oxidation of succinate to fumarate ($\Delta G^{\circ\prime} = +6 \, \text{kJ} \, \text{mol}^{-1}$). The FAD is covalently linked to the enzyme, which is tightly bound to the inner mitochondrial membrane. The electrons are passed via iron–sulphur centres to ubiquinone as part of the electron transport chain.

- *Fumarase* catalyses the addition of water to fumarate to form malate ($\Delta G^{\circ\prime} = -3.4 \, \text{kJ} \, \text{mol}^{-1}$).

- *Malate dehydrogenase* catalyses the NAD-linked oxidation of malate to regenerate oxaloacetate ($\Delta G^{\circ\prime} = +29.7 \, \text{kJ} \, \text{mol}^{-1}$). The very unfavourable standard free energy change is overcome by the very low concentrations ($<10^{-6}$ M) of oxaloacetate in the mitochondrial matrix.

Several of the intermediates of the cycle are often used for biosynthetic reactions in the cell. This immediately raises a problem since, if intermediates are removed for other purposes, oxaloacetate will not be regenerated and acetyl-CoA can only enter the cycle by condensation with oxaloacetate. The cycle therefore has to have some way of replenishing its intermediates. In animals this is done mainly by the action of *pyruvate carboxylase* (Figure 5.10). This carboxylates pyruvate to form oxaloacetate, but the energetics of the reaction require the coupling of the reaction to ATP hydrolysis. In plants and bacteria, phosphoenolpyruvate from glycolysis can be carboxylated without the involvement of ATP because of the high free energy of hydrolysis of phosphoenolpyruvate. A further reaction available to most cells is the carboxylation of pyruvate to malate. The so-called *malic enzyme* uses $NADP_{red}$ to supply the electrons for the reaction.

### 5.7.3 *Stage (iii). The electron transfer chain*

In most organisms, especially those in an aerobic environment, electron carriers are arranged to form *electron transport chains* which pass electrons in a sequence of reactions from primary reductants such as $NAD_{red}$ to final acceptors such as oxygen or nitrate (see Figure 4.5). This arrangement ensures that the free-energy yield of the overall reaction is released in a series of small steps that can easily be coupled to endergonic reactions such as ATP formation and ion transport. The free energy available in the transfer of electrons from $NAD_{red}$ to oxygen can easily be estimated from a knowledge of the redox potentials of the two compounds and the

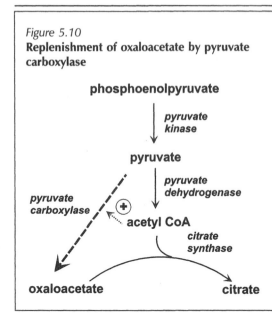

*Figure 5.10*
**Replenishment of oxaloacetate by pyruvate carboxylase**

Several of the intermediates of the citric acid cycle are used as a source of biosynthetic material. Unfortunately, if these are not replenished the cycle will be inhibited. *Pyruvate carboxylase* provides a means of producing oxaloacetate from pyruvate without first having to form acetyl-CoA and undergo the decarboxylation reactions of the citric acid cycle. The three-carbon pyruvate is carboxylated to the four-carbon oxaloacetate in a reaction that is activated by acetyl-CoA. Thus a plentiful supply of acetyl-CoA signals the need for more oxaloacetate. The carbon dioxide for the carboxylation reaction is covalently attached to *biotin*, the prosthetic group of pyruvate carboxylase

relative concentrations of their oxidised and reduced forms. Thus $E^{\circ\prime}$ for NAD is $-0.32$V. The ratio of oxidised to reduced forms of NAD in the mitochondrial matrix is around 10. Hence, from Equation (4.9) the actual potential will be $-0.29$ V. Similarly, the concentration of oxygen in the mitochondrion is much less than 1 atmosphere ($E^{\circ\prime} = +0.82$ V), being around 10–100 µM depending on the activity of the electron transport chain. A reasonable value for $E^\prime$ might be $+0.76$ V. Thus the redox gap between the entry of electrons from $NAD_{red}$ to oxygen is 1.05 V. The free energy change for the transfer of two electrons from different substrates to oxygen can be calculated from Equation (4.4):

$$\Delta G = -nF\,\Delta E$$

From $NAD_{red}$ (effective redox potential of $+0.76$ V):

$$\Delta G^\prime = -2 \times 96\,485 \times 1.05$$
$$= -203\,kJ\,mol^{-1}$$

From succinate (effective redox potential around 0 V):

$$\Delta G^\prime = -2 \times 96\,485 \times 0.76$$
$$= -147\,kJ\,mol^{-1}$$

From cytochrome $c^{2+}$ (effective redox potential around 0.3 V):

$$\Delta G^\prime = -2 \times 96\,485 \times 0.46$$
$$= -89\,kJ\,mol^{-1}$$

These reactions are coupled to ATP formation from ADP and inorganic phosphate. The standard free energy for ATP formation $(\Delta G^{\circ\prime})$ is $+31\,\text{kJ}\,\text{mol}^{-1}$. However, in the matrix of the mitochondrion the $[\text{ATP}]/[\text{ADP}][\text{P}_i]$ ratio is high, around $10^4–10^5$. This makes the effective $\Delta G'$ equal around $58\,\text{kJ}\,\text{mol}^{-1}$ (see Equation (3.3)). We can now see what the *theoretical* value for the maximum number of ATP molecules formed per pair of electrons passed down the chain could be for different substrates (the P/O ratios, see Box 5.1). These values are 3.5 for $\text{NAD}_{\text{red}}$ oxidation, 2.5 for succinate oxidation and 1.5 for cytochrome $c^{2+}$ oxidation. What are these values in practice? Most text books give values of 3, 2 and 1. There is only one reason for sticking to whole numbers and that is if the mechanism of coupling the ligand reaction to the redox reaction involves an obligatory chemical intermediate with a 1:1 stoichiometry between the redox components and ATP. No such chemical intermediate has ever been found for the electron transport chain. In fact the mechanism of coupling involves an ion gradient as an energy coupling intermediate (see Chapter 7) and stoichiometries different from whole numbers are easily accommodated. We are therefore safest sticking to experimental evidence, which gives values of around 2.6, 1.6, and 1.2 for the oxidation of $\text{NAD}_{\text{red}}$, succinate and cytochrome $c^{2+}$, respectively (Table 5.3).

**Table 5.3** P/O ratios for the oxidation of $\text{NAD}_{\text{red}}$, succinate and cytochrome $c^{2+}$

| Substrate | P/O | |
|---|---|---|
| | Maximum theoretical value | Maximum measured values |
| $\text{NAD}_{\text{red}}$ | 3.5 | $2.6 \pm 0.3$ |
| succinate | 2.5 | $1.6 \pm 0.1$ |
| cytochrome $c^{2+}$ | 1.5 | $1.2 \pm 0.2$ |

*Box 5.1* **P/O ratios**

ATP formation in mitochondria is coupled to the transfer of electrons from various substrates to oxygen. The ratio between the number of molecules of ATP formed for each oxygen *atom* reduced is known as the ATP/O ratio. Since the formation of one molecule of ATP requires one molecule of phosphate ($P_i$), the ATP/O ratio can also be stated as the P/O ratio. The definition uses a stoichiometry relative to oxygen atoms because most substrates ($\text{NAD}_{\text{red}}$, acyl-CoA, succinate, etc.) donate two electrons to the chain on oxidation and two electrons are needed to reduce one atom of oxygen to one molecule of water.

## 5.8 The energetics of glucose and fatty acid oxidation

The two principal energy substrates in mammalian tissues and in most other eukaryotic organisms are (i) glucose (stored as glycogen) (ii) fatty acids (stored as triacylglycerol (triglyceride)). Each can generate acetyl-CoA, the substrate for the tricarboxylic acid cycle. The carbons of the acetyl group are then oxidised to $CO_2$, with the electrons being used to reduce molecular oxygen by the action of the mitochondrial electron transport chain.

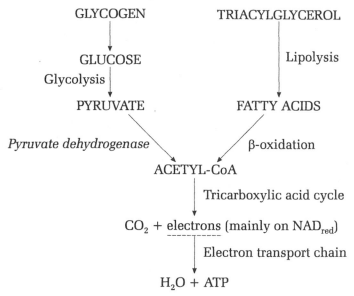

Oxidation of 1 mole of glucose to 2 moles of pyruvate yields a net production of two ATP and two $NAD_{red}$. Oxidation of pyruvate to acetyl-CoA and $CO_2$ gives another mole of $NAD_{red}$ . What happens to acetyl-CoA in one turn of the cycle? One mole of ATP is produced by substrate-level phosphorylation. The redox reactions of the cycle give 3 moles of $NAD_{red}$ and another mole equivalent pair of electrons which is directly donated by succinate to the electron transport chain. In the electron transport chain, oxidation of $NAD_{red}$ yields 2.6 moles of ATP per mole of $NAD_{red}$ and succinate oxidation provides another 1.6 moles of ATP. The overall oxidation of glucose to $CO_2$ and $H_2O$ can therefore be summarised as in Table 5.4. The final totals are given using the P/O ratios shown in Table 5.3.

Similarly, the maximum ATP yield from the oxidation of a fatty acid such as palmitate ($C_{16}$) from palmitoyl CoA to $CO_2$

and $H_2O$ can be calculated as 112.6 moles ATP per mole palmitoyl-CoA (Table 5.5). Clearly the oxidation of fatty acids has a far greater free energy yield than the oxidation of carbohydrate.

**Table 5.4** Maximum yield of ATP per mole of glucose oxidised to $CO_2$ and $H_2O$

The values are calculated using the measured P/O ratios shown in Table 5.3

| Step | Yield electron pairs | Yield ATP | Total net ATP |
| --- | --- | --- | --- |
| glucose $\rightarrow$ 2 pyruvate | 2 (2 $NAD_{red}$) | 2 | 2 |
| 2 pyruvate $\rightarrow$ 2 acetyl-CoA | 2 (2 $NAD_{red}$) | – | 2 |
| 2 acetyl-CoA $\rightarrow$ 4 $CO_2$ | 8 (6 $NAD_{red}$, 2 succinate) | 2 (GTP) | 4 |
| 10 $NAD_{red}$ $\rightarrow$ 5 $O_2$ | – | 26 | 30 |
| 2 succinate $\rightarrow$ $O_2$ | – | 3.2 | 33.2 |

**Table 5.5** Maximum yield of ATP per mole of palmitoyl-CoA oxidised to $CO_2$ and $H_2O$

The values are calculated using the measured P/O ratios shown in Table 5.3

| Step | Yield electron pairs | Yield ATP | Total net ATP |
| --- | --- | --- | --- |
| palmitoyl-CoA $\rightarrow$ 8 acetyl-CoA | 14 (7 $NAD_{red}$, 7 $FAD_{red}$) | – | – |
| 8 acetyl-CoA $\rightarrow$ 16 $CO_2$ | 32 (24 $NAD_{red}$, 8 succinate) | 8 (GTP) | 8 |
| 31 $NAD_{red}$ $\rightarrow$ 15.5 $O_2$ | – | 80.6 | 88.6 |
| 7 $FAD_{red}$ $\rightarrow$ 3.5 $O_2$ | – | 11.2 | 99.8 |
| 8 succinate $\rightarrow$ 4 $O_2$ | – | 12.8 | 112.6 |

## 5.9 The glyoxylate cycle

glyoxylate

Plants and microorganisms have the ability to metabolise the isocitrate formed in the citric acid cycle to glyoxylate and succinate. The glyoxylate can then react with acetyl-CoA to form malate which can then be coverted to oxaloacetate ready for condensation with another acetyl-CoA to form citrate (Figure 5.11). The *glyoxylate cycle* effectively cuts out the two decarboxylation stages of the citric acid cycle. The advantage to plants is that they are able to carry out the net synthesis of carbohydrate from fat. Animals, lacking the glyoxylate reactions, are unable to synthesise glucose from

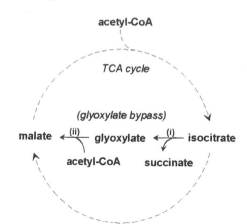

*Figure 5.11*
**The glyoxylate cycle of plants and microorganisms**

The enzymes *isocitrate lyase* (i) and *malate synthase* (ii) are present in plants and bacteria. They catalyse reactions that bypass the two decarboxylations of the citric acid cycle. Thus acetyl-CoA can be used to synthesise four-carbon units for further biosynthesis.

fatty acids since the starting material for *gluconeogenesis* (the formation of glucose from pyruvate) is pyruvate. Any acetyl-CoA formed from the oxidation of fatty acids cannot be converted back to pyruvate because of the very large positive free energy change required to reverse the pyruvate dehydrogenase step. It also cannot be used for any net synthesis of oxaloacetate, the first intermediate in gluconeogenesis (made from pyruvate by the action of pyruvate carboxylase). To form oxaloacetate from acetyl-CoA in animals requires the action of the citric acid cycle. Unfortunately, two decarboxylations take place in the cycle and hence there can be no net formation of oxaloacetate from acetyl-CoA.

Having the capability of converting fat to carbohydrate is essential for many plants which effectively store free energy in the form of triglyceride in their seed. The four major vegetable-oil crops are soybean, oil palm, rapeseed and sunflower (Figure 5.12). In germinating seeds, the reactions of the glyoxylate cycle are confined to specialised peroxisomes termed *glyoxysomes*.

Microorganisms with the glyoxylate pathway are able to grow on acetate as the sole carbon source. Some free energy has to be used in the conversion of acetate to acetyl-CoA by acetate thiokinase (Equation (5.12)), but the advantage of being able to grow on commonly available two-carbon substrates outweighs any loss in free energy yield.

$$\text{acetate} + \text{CoA} + \text{ATP} \rightarrow \text{acetyl-CoA} + \text{AMP} + \text{pyrophosphate} \tag{5.12}$$

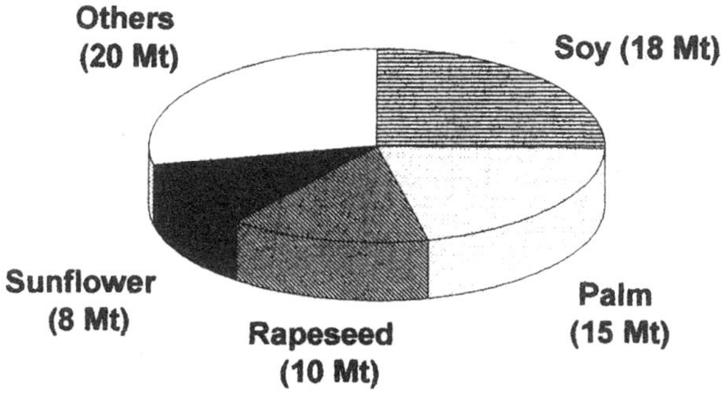

Figure 5.12
**Global plant oil production**

**Others
(20 Mt)**

**Soy (18 Mt)**

**Sunflower
(8 Mt)**

**Rapeseed
(10 Mt)**

**Palm
(15 Mt)**

The figures are averaged over the 5-year period 1993–1997. (Data from Murphy (1996).)

## Summary

- Many exergonic reactions in metabolism involve redox reactions. Changes in the oxidation state of carbon can be used to analyse metabolic reactions. The oxidation number of an atom in a compound is defined as the apparent charge the atom would have if all its bonds were ionic and the electrons were assigned to the more electronegative atoms in the compound. Thus the carbon atom in methane has an oxidation number of $-4$ (electrons from the four hydrogens assigned to the carbon), whereas in carbon dioxide it has an oxidation number of $+4$ (electrons from the carbon assigned to the two oxygens).

- Fatty acids, stored as triglyceride, provide the main energy store for animals and many plants. The oxidation of fatty acids takes place in mitochondria by a process of $\beta$-oxidation whereby the fatty acyl chains are cleaved at the $\beta$-carbon to produce acetyl fragments bound to CoA. The process involves a stepwise oxidation of the $\beta$-carbon from an initial $-2$ oxidation state to a final $+2$ oxidation state. Succesive cycles of oxidation, hydration, oxidation and thiolysis shorten the fatty acyl chains by two carbons at a time until a final acetyl-CoA is left. Odd-numbered fatty acids give rise to propionyl CoA in the final cycle.

- Fermentation is a series of catabolic reactions providing free energy without any net oxidation or reduction. However, internal redox changes in the substrates can take place. For example, the fermentation of glucose to lactate via glycolysis involves an internal redox in which two of the sugar carbons are altered to a $-3$ oxidation state and two others to a $+3$ oxidation state. The free energy of the internal redox reactions is used to drive the phosphorylation of ADP to ATP. Other fermentation pathways found in many bacteria include the Entner–Doudoroff pathway and sequences involving Stickland reactions where one compound is used as an electron acceptor and another as an electron donor.

- The pentose phosphate pathway, or hexose monophosphate shunt, for the breakdown of glucose to pyruvate produces $NADP_{red}$ and pentose sugars for the biosynthesis of nucleic acids. The reductive stages of many biosynthetic reactions, such as fatty acid biosynthesis, are specific for $NADP_{red}$ rather than $NAD_{red}$. In the cell, the ratios of $NAD_{ox}/NAD_{red}$ and $NADP_{ox}/NADP_{red}$ are not in redox equilibrium.

- Under anaerobic conditions, pyruvate can be converted to lactate. Yeasts can metabolise pyruvate to carbon dioxide and alcohol. In anaerobic microorganisms, the carbons of pyruvate can be metabolised to carbon dioxide and various reduced products such as short-chain fatty acids. The oxidation of pyruvate under aerobic conditions occurs via the citric acid cycle, which in eukaryotes takes place in mitochondria. The first stage involves the action of pyruvate dehydrogenase to release one carbon as carbon dioxide and the other two carbons as an acetyl group attached to CoA. In the citric acid cycle, or tricarboxylic acid cycle (TCA cycle), two carbons enter from acetyl-CoA and two carbons exit as carbon dioxide. The electrons from the redox reactions of pyruvate dehydrogenase and the citric acid cycle are passed to the electron transport chain via $NAD_{red}$ or directly by succinate.

- The electron transport chains of mitochondria and aerobic bacteria accept electrons from $NAD_{red}$, succinate and fatty-acyl-CoA. A series of redox carriers transport the electrons to a terminal oxidase which reduces molecular oxygen to water. The free energy from the redox reactions is coupled to the phosphorylation of ADP to ATP. The P/O ratio is defined as the ratio of the number of ATP molecules formed to the number of oxygen atoms consumed. Measured values give 2.6, 1.6 and 1.2 for the oxidation of $NAD_{red}$, succinate and reduced cytochrome $c$, respectively.

- The glyoxylate cycle of plants and microorganisms utilises some of the reactions of the citric acid cycle but bypasses the two decarboxylation stages by converting isocitrate to succinate and glyoxylate. Glyoxylate can then react with another molecule of acetyl-CoA to form malate. The advantage of these reactions to plants is that, unlike animals, they are able to synthesise carbohydrate from fat. Microorganisms can also utilise acetate for growth.

## Selected reading

Dawes, E.A., 1986, *Microbial Energetics*, London: Blackie. (Includes a comprehensive examination of microbial fermentations)

Hanson, R.W., 1990, Oxidative states of carbon as aids to understanding oxidative pathways in metabolism, *Biochem. Educ.* **18**, 194–196. (A detailed analysis of carbon oxidation states in fatty acid oxidation and the citric acid cycle)

Hinkle, P.C., 1995, Oxygen, proton and phosphate fluxes, and stoichiometries. *Bioenergetics*, Brown, G.C. and Cooper, C.E. (eds), pp. 1–16, Oxford: Oxford University Press. (A description of experimental approaches to the measurement of stoichiometries)

Kotz, J.K. and Purcell, K.F., 1991 *Chemistry and Chemical Reactivity*, Philadephia: Saunders College Publishing. (Contains a readable account of oxidation numbers and the chemistry of redox reactions)

Lehninger, A.L., 1965, *Bioenergetics: The Molecular Basis of Biological Energy Transformations*, New York: W.A. Benjamin. (One of the first, and still one of the clearest, treatments of the subject. It developed an analytical approach to bioenergetics used in many subsequent writings, even though it was written before the development of the chemiosmotic theory)

Murphy, D.J., 1996, Engineering oil production in rapeseed and other oil crops, *Trends in Biotechnol*ogy. **14**, 206–213. (Reviews progress in the use of biotechnology for oil-crop production)

Smith, C.A. & Wood, E.J., 1991, *Energy in Biological Systems*. London: Chapman & Hall. (Chapter 4, especially, gives a interesting account of the development of the citric acid cycle)

Stryer, L., 1995, *Biochemistry*, 4th edn, New York: W.H. Freeman.

Voet, D. and Voet, J.G., 1990, *Biochemistry*, New York: Wiley.

## Study problems

1.  What are the changes in oxidation state of the carbonyl carbon and the $\alpha$-carbon in the following decarboxylation reaction?

    $$RCH_2COOH \rightarrow RCH_3 + CO_2$$

2.  Using the values given in Section 5.4, calculate the ATP/(ADP)(phosphate) ratio above which no net formation of ATP can occur in glycolysis.

3.  Using the values in Section 5.4, calculate the $NAD_{ox}/NAD_{red}$ ratio below which no net formation of ATP can occur in glycolysis.

4.  Which stages of the citric acid cycle involve hydration reactions?

5.  What would be the yield of ATP per mole of acetyl-CoA oxidised via the citric acid cycle and the electron transport chain?.

6.  Why are mammals unable to convert a net amount of fatty acid into glucose?

# 6 Redox Reactions in the Biosphere

## 6.1 The oxygen cycle

Oxygen is the most abundant element on earth. It exists in the form of stable oxides with many other elements such as hydrogen, silicon and iron as well as occurring as oxygen gas. Its involvement with living systems is illustrated in Figure 1.6, where the oxidation of organic material to carbon dioxide and water is balanced by the release of molecular oxygen from water by green plant photosynthesis. The only significant production of atmospheric oxygen is by photosynthesis. Life on earth was anaerobic before the evolution of green plants and algae around 2 billion years ago (see Box 8.2).

### The kinetic stability of molecular oxygen

The standard redox potential of the oxygen couple ($O_2/H_2O$) is very positive ($+0.82$ V) and hence the reaction of most materials with oxygen takes place with a large release of energy. Fortunately for life on earth, the oxygen molecule is very stable and at room temperature is relatively inert, otherwise little free oxygen would exist in the biosphere. The kinetic barrier preventing the reaction of molecular oxygen with most other compounds arises from its electronic structure. It might be assumed at first sight that molecular oxygen would have a simple covalent double-bond structure $O{=}O$. This valence bond arrangement would seem to satisfy the pairing of the six outer electrons of the two individual atoms:

$$\ddot{\text{O}} :: \ddot{\text{O}}$$

However, the oxygen molecule shows magnetic properties, indicating the presence of two unpaired electrons, not at all obvious from the simple valence arrangement shown above. The reason for the presence of unpaired electrons can be seen from an examination of the molecular orbitals of the molecule (Figure 6.1). When the atomic orbitals at the 2p level hybridise, the resulting molecular orbitals arrange such that the $\pi^*$ orbital is at a lower energy than the $\sigma^*$ orbital.

In a triplet state, a molecule contains two unpaired electrons.

In a singlet state, all
the electrons in a
molecule are paired.

The ground state of the molecule is thus a *triplet* ($^{III}O_2$) where the spins of the two outer electrons are parallel, rather than being paired with opposing spins (*singlet state*, $^{I}O_2$). The direct reaction of a compound in a triplet state with a compound in a singlet state is very slow. There has to be a rearrangement of electron spins for such a reaction, which is energetically unfavourable. Oxygen is therefore relatively inert at room temperature. At higher temperatures, spin rearrangement is not such a problem. The electrons in both the oxygen molecule and the other reacting compound are excited to higher energy states, which can have opposing spins, and reactions to produce singlet products can then take place. The first triplet/singlet transition for molecular oxygen is at an energy level around $94\,kJ\,mol^{-1}$ above the ground state (easily reached by lighting a match!). Alternatively, singlet oxygen in this state can be generated at room temperature by excitation in the ultraviolet region of the spectrum, and in photosynthesis some singlet oxygen production occurs. Organic molecules such as carotenoids are thought to be present in photosynthetic membranes to

*Figure 6.1*

**Molecular orbital arrangement for the oxygen molecule**

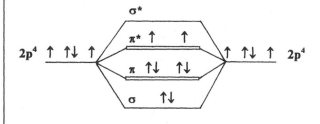

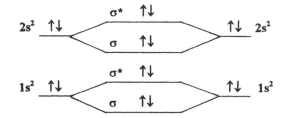

**atomic orbitals    molecular orbitals    atomic orbitals**
                    **(oxygen molecule)**

The lower energy level of the $\pi^*$ orbital compared to the $\sigma^*$ orbital means that the two outer bonding electrons in this state can maximise their spins (Hund's rule) and remain unpaired. The oxygen molecule is thus a triplet (the terminology arises from certain spectral characteristics). Energy has to be provided to rearrange the unpaired electrons into a singlet state where the spins are antiparallel.

act as quenchers of the excited oxygen molecule and prevent any damaging reactions.

### 6.1.1 *The oxidation of molecular oxygen*

The oxidation of oxygen to give ozone is energetically unfavourable ($\Delta G^{\circ} = +163$ kJ mol$^{-1}$) but in the biosphere is driven by energy from ultraviolet radiation. Photodissociation of molecular oxygen gives oxygen atoms which then react further with $O_2$:

$$O_2 \xrightarrow{\text{UV radiation}} 2O \tag{6.1}$$

$$O + O_2 \rightarrow O_3 \tag{6.2}$$

The reverse of reaction (6.2) is easily achieved by photodissociation but the oxygen atoms produced react with more $O_2$ to regenerate $O_3$. Unfortunately, the presence of certain catalysts in the atmosphere can disrupt this oxygen/ozone cycle (see Box 6.1).

Box 6.1 **The ozone layer**

The ozone layer in the earth's upper atmosphere (normally at concentrations around 5–10 ppm) plays an important role in shielding the earth from damaging ultraviolet radiation: 95–99% of the sun's ultraviolet radiation is absorbed in the oxygen/ozone cycle (Equations (6.1) and (6.2)). Certain molecules can disrupt this cycle and lead to the net breakdown of $O_3$ into $O_2$. The extra amount of high-energy ultraviolet radiation reaching the surface of the earth can lead to increased carcinogenic effects on exposed skin and damage productivity of plants and phytoplankton. Examples of compounds interfering with the oxygen/ozone cycle are chlorofluorocarbons (CFCs) produced as non-flammable, non-toxic materials for refrigerators and aerosol propellants. Chlorine radicals generated by the action of UV light on CFCs can react with $O_3$ to catalyse the net breakdown of $O_3$ into $O_2$ in an autocatalytic reaction:

$$CCl_2F_2 \xrightarrow{\text{UV radiation}} CClF_2 + Cl \tag{6.3}$$

$$Cl + O_3 \rightarrow ClO + O_2 \tag{6.4}$$

The chlorine oxide, ClO, can react with an oxygen atom, generated in the normal oxygen/ozone cycle (Equations (6.1) and (6.2)), to give $O_2$ and regenerate a chlorine radical to continue the catalytic cycle:

$$ClO + O \rightarrow Cl + O_2$$

The net reaction is the destruction of a significant amount of ozone in the atmosphere. Other compounds such as nitric oxide (NO) can interfere with the action of chlorine by reacting with ClO to release a chlorine radical and generate $NO_2$. $NO_2$ in turn can regenerate $O_3$. The effect of these complex interactions on the 35 km-deep stratosphere has been quite dramatic and a subject of concern. For the past few years there has been a measurable depletion of ozone over the Antarctic during early spring, down to less than 1 ppm at latitudes south of 70°. This 'hole' in the ozone layer appears to be increasing year by year and losses of up to 3% per annum are now seen in northern latitudes. Statutory reductions in the production and use of CFCs have now been agreed by many countries. Unfortunately, most CFCs last for over 50 years in the atmosphere and significant effects to reverse ozone depletion will take well into the next century.

### 6.1.2  *Oxygen reduction*

In a doublet state, a
molecule contains an
unpaired electron.

Two main mechanisms have evolved in living systems to
catalyse the reactions of molecular oxygen in a controlled
way. The first involves *radical intermediates.* Free radicals,
with a single unpaired electron, do not suffer from the
kinetic barrier associated with spin inversion. The products
of the reaction are radicals but the unpaired spins are now
separated and spin–spin interaction is lost. Flavin coen-
zymes are a good example of biological compounds able to
act as radical intermediates in reactions with molecular oxygen. The ring structure of the flavin (see Figure 4.1) serves as
a two-electron acceptor. However, the system can form a
stable one-electron intermediate, the semiquinone free radical. The stability of the intermediate gives flavins a catalytic
advantage over some of the other redox intermediates in
biology, for example the $NAD_{ox}/NAD_{red}$ couple, since flavins
can act as intermediates for reactions involving triplet compounds. Hence, most flavins can act as oxidases.

The second main mechanism for catalysing oxygen reduction in a controlled manner is to involve the d-orbital elec-

Box 6.2  **The transition metal chemistry of iron**

Electrons can be assigned to the five d orbitals
of transition metals according to *Hund's rule,*
which states that the most stable arrangement
is one with the maximum number of unpaired
electrons, all with the same spin orientation.
Iron, for example, with an atomic number 26,
has the electronic configuration $3d^6 4s^2$ for its
outer orbital electron:

atomic Fe

five 3d orbitals        4s

Loss of the two 4s electrons gives ferrous
iron:

ferrous $Fe^{2+}$

Further loss of electrons produces higher
valency forms:

ferric $Fe^{3+}$

ferryl $Fe^{4+}$

The versitility of a transition metal such as
iron lies in its ability to form coordination
compounds with a variety of ligands. When
coordinated in an octahedral geometry, as
found in haem, the five d orbitals are not
equivalent in energy. Two of the orbitals have
higher energy than the other three (see Figure
6.2). The energy difference $\Delta E$ between the
two sets of d orbitals depends on the electronegativity of the ligands bound to the haem.
Strongly electronegative ligands such as oxygen
and cyanide bind along the same axis as the
two higher-energy orbitals and make them difficult to fill with the iron d electrons. The
complex becomes 'low-spin' as the d electrons
are forced to pair in the lower d orbitals.
Binding of weakly electronegative ligands such
as water or fluoride lowers $\Delta E$ and the complex can then become 'high-spin'. Spin
changes in the d electrons of iron can compensate for spin changes of oxygen during
reaction and thus overcome the kinetic
restraint on the reactions of triplet oxygen
with singlet molecules.

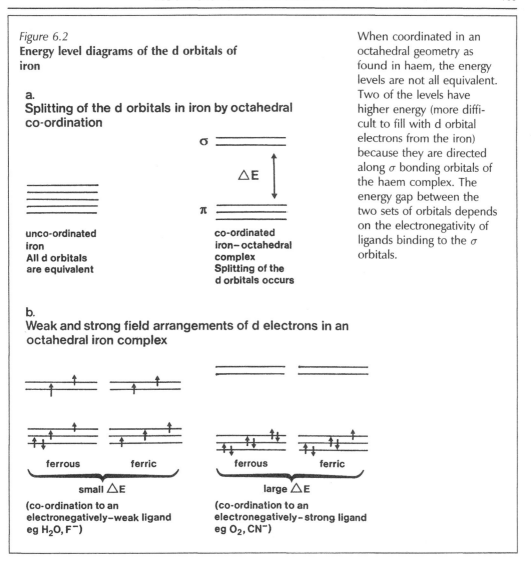

Figure 6.2
**Energy level diagrams of the d orbitals of iron**

**a.**
**Splitting of the d orbitals in iron by octahedral co-ordination**

σ

$\triangle E$

π

unco-ordinated iron
All d orbitals are equivalent

co-ordinated iron–octahedral complex
Splitting of the d orbitals occurs

When coordinated in an octahedral geometry as found in haem, the energy levels are not all equivalent. Two of the levels have higher energy (more difficult to fill with d orbital electrons from the iron) because they are directed along σ bonding orbitals of the haem complex. The energy gap between the two sets of orbitals depends on the electronegativity of ligands binding to the σ orbitals.

**b.**
**Weak and strong field arrangements of d electrons in an octahedral iron complex**

ferrous            ferric                     ferrous            ferric

small $\triangle E$                                large $\triangle E$

(co-ordination to an electronegatively–weak ligand eg $H_2O$, $F^-$)

(co-ordination to an electronegatively–strong ligand eg $O_2$, $CN^-$)

trons of certain *transition metals* such as iron (see Box 6.2). On binding molecular oxygen, changes in spin arrangement of the metal electrons in the d orbitals can take place and compensate for spin changes in the oxygen during reaction (Figure 6.2). Provided the spin state of the overall complex remains constant, then the kinetic restraint on the reactions of triplet oxygen with singlet molecules is removed. The changes in spin state of the iron come about because of the distortion of the electrostatic field of the iron by the electronegative oxygen. In haemoglobin, the distortion is coupled to conformational changes in the globin chains (Figure 6.4) which alter the affinity of the remaining oxygen-binding sites for oxygen (Box 6.3). This can be considered an example

*Box 6.3*   **Oxygen binding to haemoglobin**

Oxygen circulates around the body of vertebrates in haemoglobin, which is packed inside erythrocytes, or red cells, at high concentration. Oxygen solubility in water at 25°C is around 260 µM. By complexation to haemoglobin, the effective concentration of oxygen is raised 30-fold, thus increasing the efficiency of oxygen delivery by the circulatory system. At the relatively high oxygen concentration in the lungs, haemoglobin picks up oxygen (Figure 6.3) In the peripheral tissues, where oxygen concentrations are low, oxygen dissociates from haemoglobin. It is essential for reversible oxygen binding that the iron atom in each of the four haems of haemoglobin is kept in the reduced or ferrous form (Figure 6.4). Oxidation to the ferric form, *methaemoglobin*, not only makes the iron incapable of binding oxygen but also leads to entry of water into the haem pocket, oxidative damage to the globin chains, and eventually haemolysis.

Methaemoglobin can be re-reduced to the ferrous form by methaemoglobin reductase using $NAD_{red}$ from glycolysis. Oxidative damage is also prevented by high levels (around 5 mM) of glutathione in the red cell. *Glutathione* is an intracellular tripeptide which in the reduced form can remove any damaging hydrogen peroxide formed by the partial reduction of molecular oxygen. The pentose phosphate pathway provides $NADP_{red}$ to maintain glutathione (GSSG) in its reduced form (GSH) by the enzyme glutathione reductase:

$$GSSG + NADP_{red} = 2GSH + NADP_{ox}$$

*Glucose-6-phosphate dehydrogenase deficiency* lowers the flux rate through the pentose phosphate pathway and the intracellular concentrations of $NADP_{red}$ fall. This inherited deficiency makes the individual particularly susceptible to oxidative stress. For example, the antimalarial drug primaquine and other related compounds tend to generate damaging peroxides. In glucose-6-phosphate dehydrogenase deficiency, the inability to deal with these products leads to a build up of methaemoglobin, red cell damage and haemolysis.

*Figure 6.3*
**Oxygen binding curve for haemoglobin and myoglobin**

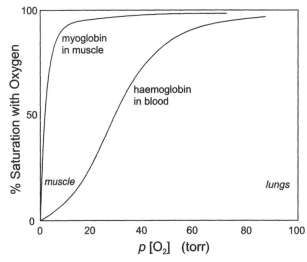

The percentage saturation of the oxygen-binding sites (four for haemoglobin and one for myoglobin) is plotted as a function of the partial pressure of oxygen. It can be seen that myoglobin is inefficient in releasing oxygen until the surrounding oxygen concentration falls to very low values.

**Glu - Cys - Gly**

The tripeptide glutathione
(The glutamic acid residue is attached to cysteine by its γ-carboxyl group)

*Figure 6.4*

**Spin changes in the d-orbitals of iron on oxygen binding**

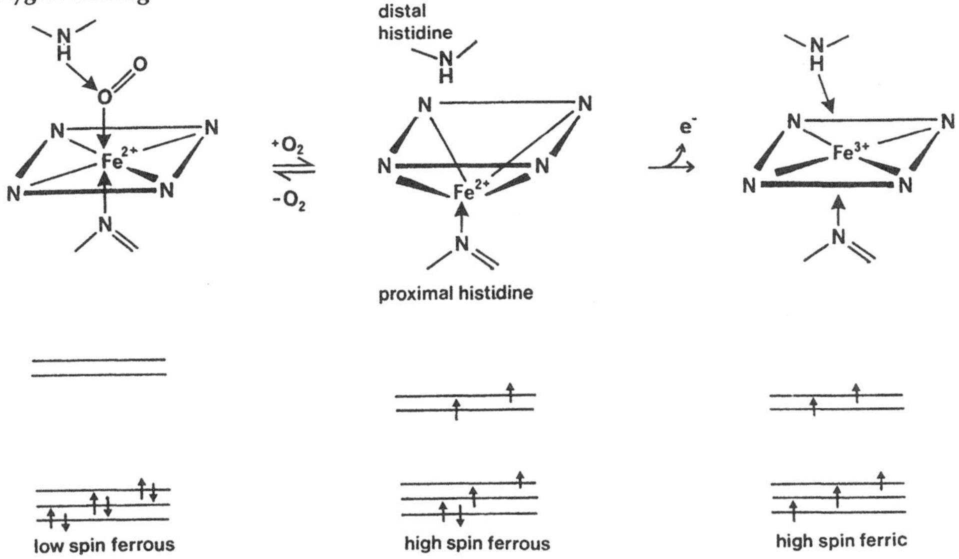

low spin ferrous      high spin ferrous      high spin ferric

Geometric and spin state changes of iron in haemoglobin (the movements of the iron are very much exaggerated). Oxygen binding to the distal position of the haem induces a small (0.7 Å) shift of the iron into the plane of the porphyrin ring and a redistribution of d electrons into a low-spin ferrous state. Oxidation of the iron to methaemoglobin forms a high-spin state which no longer binds oxygen. (From Wrigglesworth and Baum (1980).)

of mechano-chemical coupling (see Section 3.4) but, since it involves ligand binding without any accompanying chemical reaction, the energies involved are relatively small.

The energetics of oxygen reduction are fairly straightforward and the reactions are catalysed by a variety of enzymes (Table 6.1). A one-electron addition generates a superoxide radical, $O_2^-$. Controlled production of superoxide is relatively rare in biology, being generated usually by uncontrolled side-reactions of oxygen with damaged metalloproteins. The exception is in the neutrophil, where a plasma membrane oxidase produces superoxide and other radical intermediates to kill engulfed bacteria. The redox potential for the superoxide reaction is $-0.33\,V$ at pH 7, which requires a relatively strong reducing reagent, because of the negative redox potential of the $O_2/O_2^-$ couple (Figure 6.5). A more favourable reaction on thermodynamic grounds is the addition of two electrons to produce peroxide ($E^{o\prime} = +0.31\,V$), but as mentioned above this requires either a flavin or a transition metal such as iron to help overcome the kinetic restraint of adding a pair of electrons to the triplet

**Table 6.1**  Some haem-containing enzymes involved in oxygen reduction

| Reaction | Enzyme | Properties |
| --- | --- | --- |
| $O_2 + e^- \rightarrow O_2^-$ | NADP$_{red}$ oxidase | A trans-plasma-membrane complex in the neutrophil. Contains flavin and haem $b$. Used to generate toxic oxygen intermediates to kill bacteria. |
| $2O_2^- + 2H^+ \rightarrow O_2 + H_2O_2$ | superoxide dismutase | A cytosolic form contains copper and zinc; mitochondrial forms contain manganese |
| $2H_2O_2 \rightarrow 2H_2O + O_2$ | catalase | A haem protein with very high catalytic activity |
| $H_2O_2 + 2AH \rightarrow 2H_2O + 2A$ | peroxidases | A family of haem proteins catalysing the oxidation of a variety of substrates. A selenium-containing glutathione peroxidase is found in red cells |
| $O_2 + 4e^- + 4H^+ \rightarrow 2\,H_2O$ | oxidases | The terminal components of aerobic respiratory chains. Most contain copper and haem but plants also contain a non-haem iron terminal oxidase |

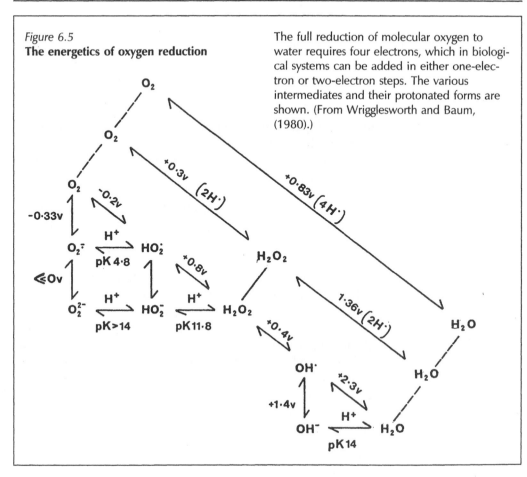

*Figure 6.5*
**The energetics of oxygen reduction**

The full reduction of molecular oxygen to water requires four electrons, which in biological systems can be added in either one-electron or two-electron steps. The various intermediates and their protonated forms are shown. (From Wrigglesworth and Baum, (1980).)

oxygen molecule. Alternatively, two one-electron steps, if coupled in sequence, would be very favourable. The formation of superoxide would be followed by a very favourable second electron addition. Superoxide is a strong oxidising agent ($E^{o'}O_2^-/H_2O_2 = +0.94\,\text{V}$) and would easily accept an electron from most compounds to form peroxide. The full reduction of peroxide to water, either by two one-electron stages (catalysed by peroxidases) or one two-electron stage (catalysed by catalase), also has a large negative free energy change. The full reduction of molecular oxygen to water is carried out by oxidases. In mitochondria the terminal complex of the electron transport chain is cytochrome oxidase, which contains three copper atoms as well as two haems (see Section 4.3). Most prokaryotes use ubiquinol as the electron donor rather than reduced cytochrome $c$.

## 6.2 The nitrogen cycle

Just about all living systems can convert ammonia ($NH_3$) to organic nitrogen (molecules containing $C-N$ bonds) but few can synthesise ammonia from atmospheric nitrogen ($N_2$) or from nitrate ($NO_3^-$), a common inorganic form of nitrogen in soil. The fixation and utilisation of nitrogen by organisms involves a series of redox reactions which form an essential component of the energetics of the biosphere (Figure 6.6).

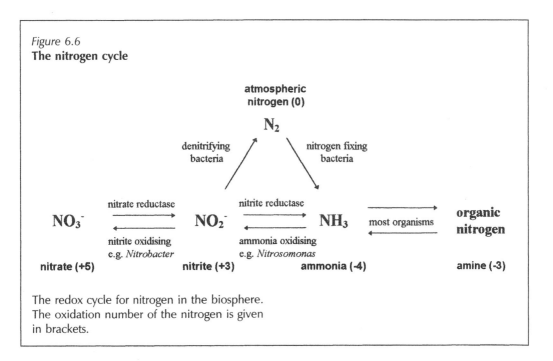

*Figure 6.6*
**The nitrogen cycle**

The redox cycle for nitrogen in the biosphere. The oxidation number of the nitrogen is given in brackets.

The reduction of $N_2$ to $NH_3$ is termed *nitrogen fixation*. Although some fixation occurs in atmospheric lightning discharges and by burning in air at high temperature, around 85% of nitrogen fixation is of biological origin. The ability to carry out the reaction is restricted to certain bacteria and cyanobacteria. No eukaryotic organisms can fix nitrogen.

### 6.2.1  Nitrogen fixation from atmospheric nitrogen

The stoichiometry of nitrogen fixation by *nitrogenase* involves a progressive six-electron reduction:

$$N_2 + 6e^- + 6H^+ \rightarrow 2NH_3 \qquad (6.5)$$

The reaction is slightly exothermic ($\Delta G^{\circ\prime} = -33\,\text{kJ}\,\text{mol}^{-1}$) but unfortunately the triple bond of nitrogen, $N\equiv N$, is extremely stable. It is necessary to overcome a large energy of activation by coupling the reaction to the free energy change from the hydrolysis of at least 16 ATP molecules (the precise stoichiometry is not yet known).

$$N_2 + 8e^- + 10H^+ + 16ATP \rightarrow 2NH_3 + H_2 + 16ADP + 16P_i$$
$$(6.6)$$

Nitrogen fixation is one of the most energy-expensive reactions in biology (compare the four electrons and three ATP molecules required to fix each $CO_2$ in the Calvin cycle of photosynthesis described in Chapter 8). Few organisms, only some bacteria and cyanobacteria, have this ability. One goal of plant genetics and molecular biology is to incorporate the nitrogenase system into non-leguminous plants to cut down the use of nitrogen fertilizers in crop production.

A central problem for nitrogen fixers is that the nitrogenase enzyme complex is very sensitive to oxygen. Anaerobic organisms such as *Clostridium* and *Desulfovibrio* are protected because of their oxygen-free environment, but aerobic organisms have had to evolve protective mechanisms. The aerobic soil-bacteria *Klebsiella* and *Azotobacter* are thought to possess very active oxidases which help to maintain a low oxygen concentration around the nitrogenase complex. The bacteria of the genus *Rhizobium* live on the roots of legumes such as peas and beans. The root nodules contain leghaemoglobin, an oxygen-binding protein which also helps to maintain a low-oxygen environment (Figure 6.7).

Cyanobacteria have evolved to localize the nitrogenase in specialized cells called heterocysts, which have thickened

*Figure 6.7*
**The root nodule of soybean**

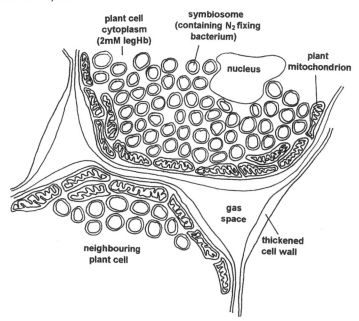

The plant cell cytoplasm is packed with symbiosomes, membrane-enclosed nitogen-fixing bacteria. The oxygen concentration in the cytoplasm is maintained at around 50 nM by the presence of leghaemoglobin at high concentration and the activity of the plant mitochondria. The cell wall adjacent to the gas spaces between cells is thickened to reduce the rate of oxygen diffusion into the cell.

cell walls to lower the diffusion of oxygen into the sensitive nitrogenase complex.

The *nitrogenase complex* comprises two components:

Component I    Dinitrogenase which catalyses the reduction of $N_2$. An $\alpha_2\beta_2$ tetramer of 240 kDa containing two iron–sulphur clusters and an iron/molybdenum cofactor which serves as the site for $N_2$ binding.

Component II   Nitrogenase reductase (component II), a dimer of identical 30 kDa subunits, contains an iron–sulphur cluster which provides the electrons for component I.

The sequence of electron flow from reducing metabolites to $N_2$ differs slightly according to the organism, but component II collects electrons from substrates such as pyruvate via ferredoxin (in *Rhizobium*) or flavodoxin (in *Klebsiella*) (Figure 6.8). Component II then binds ATP, causing a conformational

*Figure 6.8*
**Electron flow through the nitrogenase complex of *Klebsiella***

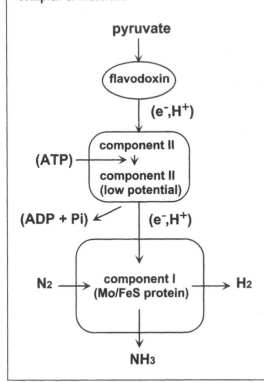

Electrons and protons are transferred from pyruvate to component I of the nitrogenase complex via flavodoxin and component II. ATP acts to alter the conformation of component II to a low redox potential state. The overall stoichiometry requires 8 electrons and 8 protons, and at least 16 ATPs. Molecular hydrogen is an obligatory product of the reaction.

change in the protein and a lowering of the redox potential from around $-0.29$ V to $-0.4$ V. Transfer of electrons to component I then takes place. Nitrogen gas binds to the molybdenum cofactor in component I (releasing bound hydrogen in the form of $H_2$ in the process) and electron transfer continues until the six-electron reduction of dinitrogen occurs. In anaerobes, the released hydrogen gas can be recycled and act as an extra electron donor in addition to pyruvate.

### 6.2.2 *Nitrogen fixation from nitrate*

The other pathway for the formation of ammonia from nitrate and nitrite (the *assimilatory pathway*) is much more common and can be carried out by most plants and microorganisms. The amount of nitrogen fixed via the two-step reduction of nitrate to ammonia exceeds that fixed by nitrogenase by over 100-fold. A large proportion of terrestrial nitrogen fixation is carried out by agricultural crops stimulated by the addition of nitrogen fertilizers. Unfortunately, although fertilisers are essential for modern food production, their overuse can lead to contamination of groundwaters. This is a cause for concern since nitrite can be converted to nitrosamines, potential carcinogenic compounds. The denitrification of excess soil nitrites by bacteria can also produce nitrous oxides, contributing to air polution.

The first stage of the reaction sequence is the reduction of nitrate to nitrite by the *assimilatory nitrate reductase*, a protein complex containing FAD, molybdenum and cytochrome *b*. The initial electron source is either $NAD_{red}$ (higher plants and algae) or $NADP_{red}$ (fungi and bacteria). Electrons reach the molybdenum cofactor via FAD and cytochrome *b*.

$$2e^- + NO_3^- + 2H^+ \rightarrow NO_2^- + H_2O \qquad (6.7)$$

The further reduction of nitrite to ammonia is carried out by the *assimilatory nitrite reductase* a soluble enzyme containing an iron–sulphur centre and *sirohaem* (haem structures in which the porphyrin ring is partially reduced). The six-electron reduction of nitrite to ammonia involves reduced ferredoxin as the electron donor in higher plants and $NAD(P)_{red}$ in fungi and bacteria. The three redox stages of the overall reaction

$$NO_2^- \xrightarrow{2e^-} [NO^-] \xrightarrow{2e^-} NH_2OH \xrightarrow{2e^-} NH_3$$

nitrite    [nitroxyl]    hydroxylamine    ammonia

are all carried out by the same enzyme. There is no conservation of free energy in this reaction.

Once ammonia has been synthesised, it can be assimilated into organic compounds via reactions leading to glutamate, glutamine and carbamoyl phosphate, reactions common to all organisms.

### 6.2.3 *Oxidation of ammonia*

The reverse process, the oxidation of ammonia to nitrite and nitrate, is limited to bacteria and uses molecular oxygen as the terminal electron acceptor. However, no single bacterium can carry out both reactions. The full oxidation of ammonia results from the sequential action of two groups of bacteria, the ammonia-oxidising group (an example is *Nitrosomonas*) and the nitrite-oxidising group (an example is *Nitrobacter*). Both organisms are known as *lithotrophs*, being able to derive their energy from oxidising inorganic compounds.

The oxidation of ammonia to nitrite is overall exothermic ($\Delta G^{\circ\prime} = -272 \, kJ \, mol^{-1}$) but appears to involve the unfavourable formation of hydroxylamine ($NH_2OH$) as an intermediate. This is a strong oxidising agent ($E^{\circ\prime}$ of the ammonia/

hydroxylamine couple is +0.9 V), which would make it difficult for the reaction to pass electrons directly to the electron transport chain. However, the second stage to form nitrite has a much lower redox potential (+0.06 V) and hence the overall reaction can use oxygen as the final electron acceptor and derive free energy from coupling ATP formation to the electron transfer.

In the nitrite oxidase system of *Nitrobacter*, the electrons from nitrite are transported to oxygen via an electron transport chain to generate ATP. The redox potential of the nitrite/nitrate couple is quite positive (+0.42 V) compared to the oxygen/water couple (+0.82 V) and the free energy change is small, resulting in low growth yields for *Nitrobacter*. The reaction requires an energy-dependent reduction of cytochrome *c* (+0.25 V) before nitrite reduction can be coupled to ATP formation. A chemiosmotic loop provides a simple mechanism (see Chapter 7).

### 6.2.4 *Denitrification*

A second route for the reduction of nitrate to nitrite followed by further reduction of the nitrite to $N_2$ can only be carried out by various facultative anaerobes (the *dissimilatory pathway* for nitrate and nitrite reduction). It should be noted that although the same names are used for nitrate and nitrite reductases in the assimilatory and the dissimilatory pathways, these are separate, distinct enzyme systems. Nitrate is used as an electron acceptor in place of oxygen by the membrane-bound nitrate reductase reaction (dissimilatory). Electron donors vary but can include $H_2$. The transfer of reducing equivalents to the nitrate/nitrite couple (+0.42 V) involves transfer of charge across the membrane and the free energy from the reaction can be coupled to ATP production. The nitrite produced can then be converted by the same bacteria to $N_2$ in a two-step process. A soluble nitrite reductase first converts nitrite to nitric oxide:

$$NO_2^- + 2H^+ + e^- \rightarrow NO + H_2O \qquad (6.8)$$

A transmembrane complex, nitric oxide reductase, then converts nitric oxide to nitrous oxide in a free-energy conserving step:

$$2NO + 2H^+ \rightarrow N_2O + H_2O \qquad (6.9)$$

Finally a copper-containing nitrous oxide reductase generates molecular nitrogen:

$$N_2O + 2H^+ + 2e^- \rightarrow N_2 + H_2O \qquad (6.10)$$

All four reactions, from nitrate to molecular nitrogen, are closely linked and the overall denitrification process (dissimilatory) conserves free energy by coupling the electron flow to ATP production via a chemiosmotic mechanism (Chapter 7).

## 6.3 The sulphur cycle

Sulphur can take up a variety of oxidation states in the biosphere (Figure 6.9). It can exist as hydrogen sulphide ($H_2S$, oxidation state $-2$), sulphur dioxide ($SO_2$, oxidation state $+4$), sulphite ($SO_3^{2-}$, oxidation state also $+4$), sulphate ($SO_4^{2-}$ oxidation state $+6$), as well as elemental sulphur ($S^0$). It has a high affinity for metals and iron–sulphur clusters are components of many electron transport chains. Organic sulphur is essential for life, being incorporated into many proteins as cysteine and methionine as well as forming thioesters, for example in acetyl-CoA.

A major component of atmospheric sulphur is sulphur dioxide, produced mainly by human activity. It is estimated that over 200 million tonnes of sulphur are released into the atmosphere each year by the roasting of sulphide ores and the combustion of sulphur-containing coal and fuel oil. This is more than half of the total sulphur released by all other natural sources. Wet precipitation of sulphur dioxide together with carbon dioxide (acid rain) contributes to the decline of large areas of forest and the acidification of many rivers and lakes. The other main sulphur components of the atmosphere are hydrogen sulphide, formed primarily from

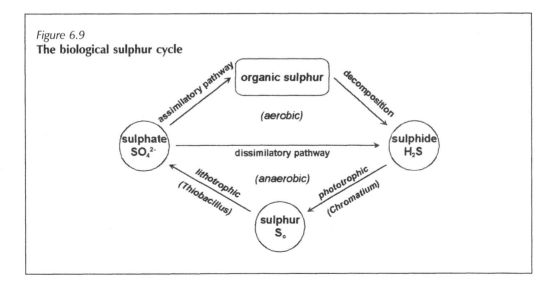

*Figure 6.9*
**The biological sulphur cycle**

the bacterial reduction of sulphate, and dimethyl sulphide ($(CH_3)_2S$) released in the decomposition of organic sulphur by marine algae. With an oxidation state of $+6$, sulphate is the most stable form of sulphur in water and soil. By far the bulk of sulphur on earth is found in sediments and rocks.

### 6.3.1 Sulphate reduction

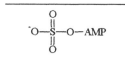

adenosine 5-phosphosulphate

Various bacteria are responsible for sulphate reduction, either to provide a source of reduced sulphur for incorporation into organic compounds (assimilatory reduction) or to use sulphate as a terminal electron acceptor in the absence of oxygen (dissimilatory reduction) (Figure 6.9). The initial problem for either pathway is the initial reduction of sulphate to sulphite. The redox potential of the sulphate/sulphite couple is very negative ($-0.52$ V), which makes sulphate a very difficult substance to reduce. In order to overcome this redox stability, ATP first reacts with sulphate to form *adenosine phosphosulphate* (APS) an activated form of sulphate ($\Delta G^{o\prime}$ for hydrolysis is $-88$ kJ mol$^{-1}$).

$$\text{ATP} + \text{SO}_4^{2-} \xrightarrow{\text{ATP sulphurylase}} \text{APS} + \text{pyrophosphate} \qquad (6.11)$$

The redox potential of the bound sulphate is effectively raised to around 0 V at the expense of free energy from the ATP reaction (the pyrophosphate is subsequently hydrolysed to two molecules of phosphate, thereby ensuring the direction of the reaction to APS formation). The activation effectively allows the sulphate to be reduced to sulphite by the action of *APS reductase*. A wide variety of aerobic organisms can carry out the subsequent six-electron reduction of sulphite to sulphide. The reaction is catalysed by *sulphite reductase*, a single enzyme containing an iron–sulphur cluster and a haem in close proximity. Sulphite reductases of anaerobes (catalysing dissimilatory reduction) are usually of lower molecular mass but have the same prosthetic groups as the aerobic enzyme. The availability of organic carbon to provide the electrons for reduction often limits the dissimilatory pathway, although non-organic substrates such as formate and hydrogen can be used. In marine sediments and wetlands, dissimilatory sulphate reduction is the main form of respiration.

### 6.3.2 Sulphide and sulphur oxidation

Since hydrogen sulphide rapidly oxidises in the presence of oxygen and transition metal catalysts, the biological oxidation of sulphide to elemental sulphur is usually catalysed under low or zero concentrations of oxygen. Photosynthetic bacteria can use sulphide as a source of electrons for carbon fixation under anaerobic conditions (see Chapter 8). The more redox-stable $S^0$ can be oxidised under aerobic conditions mainly by the genus *Thiobacillus* which produces sulphate with an accompanying acidification. This can have unfortunate environmental consequences. For example, *Thiobacillus ferrooxidans* acts on the mineral pyrite, which is a complex of iron and sulphur found in many coals and ores. The oxidising action of the bacterium produces acidic mine waters, an intractable problem for many mining companies.

## 6.4 The hydrogen cycle

Most molecular hydrogen released to the atmosphere is lost slowly by diffusion to space. However, local hydrogen cycling does take place in confined habitats. Molecular hydrogen is a common end product of many anaerobes, especially the genus *Clostridium*, that depend on fermentation for growth (Chapter 5). Pyruvate can be converted to $CO_2$, acetyl-CoA and $H_2$ (see Equation (5.7)). Ethanol can be converted to acetate and $H_2$.

Hydrogen-oxidising organisms are generally lithotrophic (see Box 6.4). For example, *methanogens* (classified as Archaebacteria) can grow solely by reducing carbon dioxide to methane using molecular hydrogen:

$$CO_2 + 4H_2 \rightarrow CH_4 + 2H_2O \qquad (\Delta G^{\circ\prime} = -131\,\text{kJ}\,\text{mol}^{-1})$$

$$(6.12)$$

They possess a nickel-containg hydrogenase which catalyses the oxidation of molecular hydrogen, with the electrons being transferred to carbon dioxide via a sequence of electron transport components. The hydrogen is supplied by hydrogenases. Some ADP is phosphorylated during electron transfer, but the yield is low since the concentration of one of the substrates, molecular hydrogen, is in the micromolar range. The biogenic production of methane is responsible for around 80% of atmospheric methane and is a significant contributor to the greenhouse effect (see Box 8.1).

The *acetogenic bacteria* are also able to use hydrogenase to supply reducing equivalents for carbon dioxide fixation, but in this case the end product is acetate rather than methane.

Figure 6.10
**The redox scale for life**

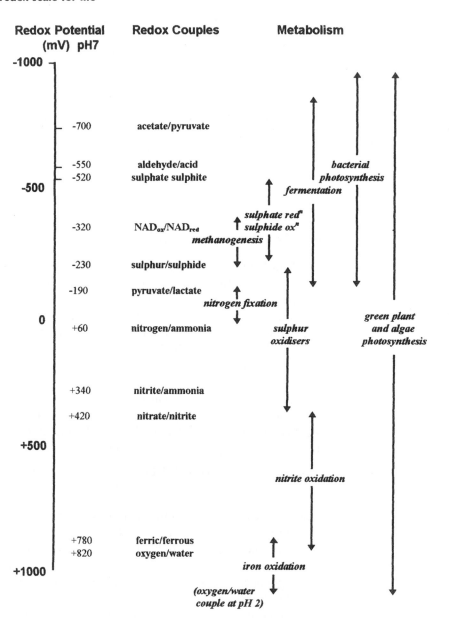

Different organisms derive their energy requirements from redox reactions between a variety of redox couples. The actual redox potential will vary from the standard value depending on the concentrations of oxidant and reductant. An extreme example is the oxidation of ferrous iron by *Thiobacillus ferrooxidans* which exploits a standard potential redox gap of only 0.04 V.

*Box 6.4* **Redox habitats**

The present atmosphere exists at an ambient redox potential of around +0.3 V. Leave a solution of reduced cytochrome *c* exposed to air and in time it will equilibrate to around 90–95% oxidised. Aerobes are able to take advantage of more positive redox potentials by direct electron transfer to molecular oxygen. However, many habitats in the biosphere are anoxic or close to anoxic and at a much more negative potential. Marine sediments provide anoxic conditions for a wide variety of organisms, as do the sediments in many lake bottoms. Oxygen diffusion is too slow over distances more than a few centimetres to be able to replenish oxygen loss due to aerobic metabolism. Deep lake waters that do not mix thermally with the upper layers can become anoxic. The largest example is the Black Sea, where oxygen does not penetrate below depths of around 150 m.

Different organisms are presented with different problems of energy transduction depending on the redox range available. *Heterotrophs* derive their energy from the catabolism of organic carbon substrates which provide reducing equivalents at potentials down to −0.5 V (Figure 6.10). With oxygen as a terminal electron acceptor, the free energy available from a redox gap of around 1.3 V makes heterotrophic aerobes very growth efficient. Heterotrophic anaerobes have less redox energy available. Organic debris accumulating from decomposed aerobes is the main source of substrate. Substrate-level phosphorylation from fermentation provides the necessary energy for growth up to redox potentials around −0.2 V. Electron transport releases more free energy by oxidising the products of fermentation using inorganic compounds such as sulphate and nitrate as electron acceptors (redox potentials up to +0.4 V). When these are absent, reducing equivalents can be transferred to methanogens, which have to exist on a redox gap of less than 0.5 V. The suggestion has been made that the reason why so few eukaryotes inhabit anaerobic environments is that the biomass production is too low to support a predatory food chain. *Autotrophs* are organisms that are able to obtain the carbon for growth from carbon dioxide. The necessary energy can come from light (*phototrophs)* or from the oxidation of inorganic compounds (*lithotrophs).* Phototrophs include green plants and algae, which use the largest redox range of all living organisms—differences of up to 1.6 V—and photosynthetic bacteria with lower but still substantial redox differences in their photosynthetic mechanisms (see Chapter 8). Lithotrophs, on the other hand, are very restricted in the availability of free energy, exploiting small redox gaps between oxygen $(E^{o\prime} = +0.82\,V)$ and reductants such as nitrite and even ferrous iron $(E^{o\prime} = +0.78\,V)$. This group of organisms includes no eukaryotes.

# Summary

- The very positive redox potential of the oxygen/water couple means that oxygen can react with most materials in the biosphere with a large release of energy. However, because it is a triplet molecule with two unpaired electrons there is a kinetic barrier inhibiting most reactions. The inhibition can be overcome by excitation of one of the electrons to a higher energy state with an opposing spin. Alternatively, free-radical intermediates such as semiquinones and flavins can be used. Binding to transition metals such as iron or copper also aids reaction as changes in spin arrangement of the metal electrons can compensate for spin changes of the oxygen during reaction.

- Most living organisms can convert ammonia to organic nitrogen but few can synthesise ammonia from atmospheric nitrogen. Nitrogen fixation by bacteria involves the progressive six-electron reduction of nitrogen and is catalysed by nitrogenase. The nitrogenase complex is sensitive to oxygen and has to be protected by mechanisms to lower the local oxygen concentration. Nitrogen fixation from nitrate and nitrite to make ammonia for organic synthesis (the assimilatory pathway) is much more common and can be carried out by most plants and microorganisms. Reduction of nitrate and nitrite to ammonia for conversion back into nitrogen gas (the dissimilatory pathway) is carried out by various facultative anaerobes.

• Organic sulphur is essential for life, being present in many proteins and metabolites. Sulphate reduction proceeds via the formation of adenosine phosphosulphate. In this activated form it can be reduced to sulphite and then sulphide by various organisms. Photosynthetic bacteria can use sulphide as a source of electrons for carbon fixation under anaerobic conditions.

• Molecular hydrogen is produced as an end product of fermentation by many anaerobes. In turn, it can be oxidised by methanogens for the production of methane from carbon dioxide.

## Selected reading

Brock, T.D., Smith, D.W. and Madigan, M.T., 1984, *The Biology of Microorganisms*, London: Prentice-Hall International. (Comprehensively covers the subject)

Fenchel, T. and Finlay, B.J., 1994, The evolution of life without oxygen, *Am. Sci.* **82**, 22–29. (A stimulating article on the evolution of anaerobic pathways)

Kelly, D.P. and Smith, N.A., 1990, Organic sulphur compounds in the environment: Biogeochemistry, microbiology and ecological aspects, *Advances in Microbial Ecology*, Vol. II, Marshall K.C. (ed.), pp. 345–385, New York: Plenum.

Williams, R.J.P. and FraústodaSilva, J.J.R., 1996, *The Natural Selection of the Chemical Elements*, Oxford: Clarendon Press. (A comprehensive and innovative treatment of the inorganic chemistry of the biosphere.)

Wrigglesworth, J.M. and Baum, H., 1980, The biochemical functions of iron, *Iron in Biochemistry and Medicine*, vol. II, Jacobs, A. and Worwood M. (eds), New York: Academic Press.

# 7 Membrane Transport: Ion Gradients as Common Intermediates

In Chapter 1, it was mentioned that work, or free energy, can be thought of as coherent or directed motion. On this basis, a gradient of molecules diffusing from high to low concentration can be made to do work (to move a piston, for example). To create a gradient requires work (we need to push back on the piston to concentrate the molecules into a smaller volume). Similarly, to create a concentration gradient across a biological membrane requires work. This can be provided by coupling the vectorial movement to other more favourable changes such as a redox reaction or the hydrolysis of ATP. Discharging the gradient releases the stored free energy, which in turn may be coupled to drive other reactions.

## 7.1 Energy in gradients

The free energy stored in a concentration gradient of an *uncharged* solute can be expressed as

$$\Delta G = -RT \ln \frac{[C_2]}{[C_1]} \tag{7.1}$$

where $[C_2]$ and $[C_1]$ represent the solute concentrations in the two compartments. $[C_2]$ is the higher concentration. This makes the ratio of $[C_2]/[C_1]$ greater than unity and hence the logarithm of the ratio will be positive. Movement of the solute from $C_2$ to $C_1$ results in a negative $\Delta G$ (favourable reaction). The expression can be compared to that for a chemical reaction (Equation (3.3)). The derivation is essentially the same.

With ions, we have to take into account the movement of charge. The equivalent expression to Equation (7.1) for charged molecules is

$$\Delta G = -RT \ln \frac{[C_2]}{[C_1]} - nF\Delta\psi \tag{7.2}$$

*Box 7.1* **A calculation example**

**Question**

Calculate the free energy change when 1 mole of glucose is transported down a glucose concentration gradient of 1000:1 at 37°C.

**Answer**

$$\Delta G = -8.314 \times 310 \times \ln(1000)$$
$$= -17\,804\,\mathrm{J\,mol^{-1}}$$

(For those familiar with logarithms to base 10, Equation (7.1) can also be written as

$$\Delta G = -2.303\,RT\,\log\frac{[C_2]}{[C_1]}$$

This makes mental calculations easier, since for every ×10 change in concentration, the free energy will change by 5936 J mol$^{-1}$ (or approximately 6000 J mol$^{-1}$).)

where $n$ is the charge on the ion, $F$ is the Faraday constant ($96\,485\,\mathrm{J\,V^{-1}\,mol^{-1}}$) and $\Delta\psi$ is the membrane potential. We have to be careful with signs at this point. As written in (7.2), the membrane potential is such that it has the same sign as the charge of the ion in the higher concentration compartment. The movement of a sodium ion ($Na^+$) across a membrane to increase the concentration of the ion on the receiving side will increase the positive potential on that side. Then the movement of $Na^+$ back across the membrane will result in a negative free energy change from the concentration term $(-RT\ln([C_2]/[C_1]))$ and the electrical term $(-nF\,\Delta\psi)$. Both the concentration gradient and the membrane potential will favour movement of the ion across the membrane.

There are many instances where the concentration and the electrical terms oppose each other. Consider the example of a semipermeable membrane which is fully permeable to cations but impermeable to anions. Let there be a greater concentration of sodium chloride on one side compared with the other. Initially $Na^+$ will move through the membrane down its concentration gradient. $Cl^-$ will not be able to pass through the membrane:

| high<br>concentration | $Na^+$ ⟶<br>$Cl^-$ | low<br>concentration |
|---|---|---|

The movement of $Na^+$ will carry positive charge across the membrane and lead to a membrane potential, positive on the low concentration side:

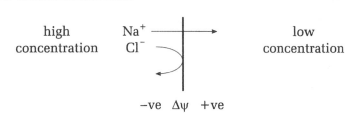

Eventually the build-up of positive charge will inhibit the further movement of $Na^+$. At this point no change occurs in the system ($\Delta G = 0$). We then have from Equation (7.2),

$$0 = -RT \ln \frac{[C_2]}{[C_1]} - nF \Delta \psi \qquad (7.3)$$

or

$$\Delta \psi = -\left(\frac{RT}{nF}\right) \ln \frac{[C_2]}{[C_1]} \qquad (7.4)$$

This equation quantifies the equivalence of concentration and electrical terms with regard to free energy change (Table 7.1). It can easily be calculated that a 10-fold concentration gradient of monovalent ions can be maintained across a membrane by an opposing membrane potential of around 60 mV (the exact value will depend on the temperature of the system). For divalent ions ($n = 2$) the corresponding potential is 30 mV. It is important to realise that the number of ions in the concentration gradient does not represent the number of charges creating the membrane potential. Only a few ions have to move down the gradient to build up sufficient opposing potential to inhibit further movement (Box 7.2).

**Table 7.1** Equivalence of concentration and electrical terms in ion gradients

Membrane potential values necessary to oppose the movement of ions down a concentration gradient. (Calculated from Equation (7.4) using a temperature of 27°C. The values would have to be adjusted slightly for other temperatures.)

| Concentration ratio $[C_2]/[C_1]$ | $\Delta \psi$ (monovalent ions) (mV) | $\Delta \psi$ (divalent ions) (mV) |
|---|---|---|
| 10 | 60 | 30 |
| 100 | 120 | 60 |
| 1000 | 180 | 90 |
| 10 000 | 240 | 120 |
| 100 000 | 300 | 150 |

*Box 7.2* **The amount of charge movement needed to create a membrane potential**

**Question**

How much charge has to be moved to create a membrane potential of 100 mV?

**Answer**

The relationship between charge ($q$) and potential ($V$) for a capacitor is

$$C = \frac{q}{V}$$

where $C$ is the capacitance measured in units of farads (F). For a 1 cm$^2$ area of membrane with a potential 0.1 V, the charge needed will be

$$q = 0.1C$$

Biological membranes have a capacitance value around 1 μF, so the charge $q$, in coulombs (C), will be $0.1 \times 10^{-6}$. 1 mole of charge carries 96 485 coulombs (the Faraday constant). Hence $0.1 \times 10^{-6}$ coulombs will be equivalent to approximately $10^{-12}$ moles of charge. In other words, it only requires the uncompensated movement of 1 picomole of ions to create a membrane potential of 100 mV across the membrane. Very little charge is taken from one compartment to the other. The imbalance of charge in terms of numbers of ions is so small as to maintain the *principle of electroneutrality* in the bulk solution. The gradient of the ion can be quite large (Table 7.1) but the compensating counterion in the compartment, chloride for example, maintains electroneutrality.

## 7.2 Passive transport

Consider the diffusion of an *uncharged* molecule across a membrane down a concentration gradient from an outside concentration $C_o$ to an inside concentration $C_i$:

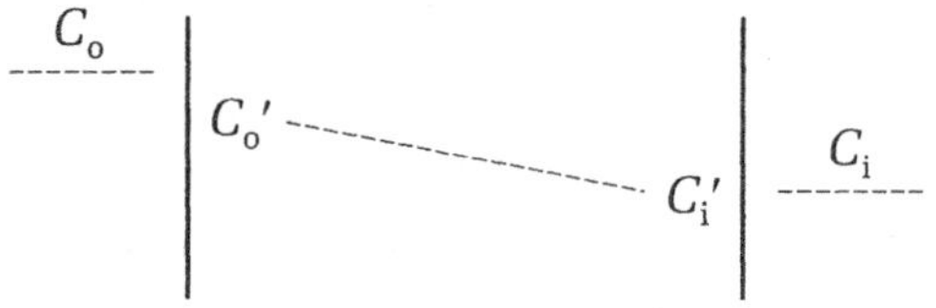

The concentrations of the molecule inside the membrane $C_o'$ and $C_i'$ will not necessarily be the same as the concentrations in the bulk solution. They will be related by the ability of the molecule to dissolve in the membrane material, which can be expressed in terms of the partition coefficient ($h$):

$$h = \frac{C'}{C} \tag{7.5}$$

Assuming that there is no build-up of molecules in the membrane, the rate of transport of molecules (the flux, $J$) that are passing through an area of membrane ($A$) at steady state can be expressed by *Fick's first law of diffusion:*

$$J = -DA\frac{\mathrm{d}C'}{\mathrm{d}x} \tag{7.6}$$

*Box 7.3* **The speed of diffusion**

Fick's law can also be applied to the random motion of molecules even in the absence of a concentration gradient. It can be shown that the mean-square displacement $x^2$ is related to the diffusion coefficient by the equation

$$x^2 = 2Dt$$

where $t$ is time. In other words, a molecule with a diffusion coefficient $D$ moves an average linear distance $x$ in a time $t$ of $\sqrt{2Dt}$. For diffusion on a surface, the right-hand side of the equation is $4Dt$, and for diffusion in three dimensions, it is $6Dt$.

The flux is proportional to the concentration gradient in the membrane ($dC'/dx$) and the diffusion coefficient ($D$) of the molecule in the membrane. The minus sign occurs because the transport is down the concentration gradient. Replacing $C'$ by $hC$ (Equation (7.5)) and integrating across the thickness of the membrane ($d$) gives

$$Jd = -DAh\,\Delta C \tag{7.7}$$

where $\Delta C$ is the concentration difference ($C_o - C_i$) across the membrane. $Dh/d$ is sometimes defined as the *permeability coefficient* ($P$) of the molecule, which then gives

$$J = -PA\,\Delta C \tag{7.8}$$

The predictions from Equation (7.8) are that the flux per unit area of uncharged molecules across a biological membrane by passive diffusion will be proportional to the concentration difference; proportional to the lipid solubility of molecule (because of $h$, the partition coefficient); and inversely proportional to the size of the molecule (because $D$ is inversely proportional to size (Table 7.2)).

**Table 7.2** Values of the diffusion coefficient for some typical molecules in water at 20°C

| Molecule | Molecular mass (kDa) | $D \times 10^{11}$ (m² s⁻¹) |
|---|---|---|
| $O_2$ | 16 | 200 |
| glycine | 75 | 106 |
| cytochrome $c$ | 12 500 | 13 |
| myoglobin | 17 800 | 11.3 |
| haemoglobin | 68 000 | 6.02 |
| urease | 480 000 | 3.5 |
| tobacco mosaic virus | 40 000 000 | 0.53 |

### 7.2.1 *Transport of charged molecules*

The partition coefficient for charged molecules is *very* small (values of around $10^{-10}$). This is because the ions are associated with surrounding water molecules whose alignment serves to lower the electrostatic potential of the charge. Removal of the hydration shell from the ion requires a large energy expenditure. However, it is possible to replace the water dipoles by other polar molecules containing dipolar atoms such as oxygen or nitrogen. This happens in the chelation of cations by certain antibiotics, for example valinomycin (see Figure 7.1). Transmembrane channels of protein can be lined with polar groups to enable the passage of ions (Figure 7.2). The ion is able to release most, or all, of its solvation shell by interacting with the dipoles on the protein. Some selectivity can be based on the size of the unhydrated ion. The sodium ion has an ionic radius of 0.95 Å compared with 1.33 Å for potassium. A sodium-specific channel can therefore select for sodium on the basis of unhydrated size. Obviously other factors have to come into play for a potassium-selective channel. In this case the geometry of the polar groups lining the channel is arranged so that replacement of the potassium hydration shell is facilitated, but the oxygen dipoles on the protein are too far apart to replace the water associated with any sodium ion. It has to stay hydrated and as such is too large to pass through.

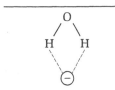

Water interaction with an anion

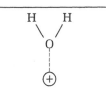

Water interaction with a cation

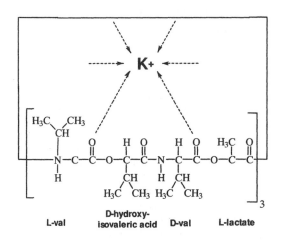

*Figure 7.1*
**Chelation of potassium by valinomycin**

L-val | D-hydroxy-isovaleric acid | D-val | L-lactate

Valinomycin is a cyclic peptide with both D- and L-amino acids linked by oxygen-ester as well as peptide bonds. It has an exterior apolar surface and a central cavity lined with polar oxygen groups. These replace the hydration shell of potassium which can then bind and be transported across the membrane. The complex retains the positive charge of the bound ion and is thus sensitive to membrane potential.

*Figure 7.2*
**Ion transport through transmembrane channels**

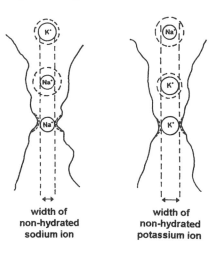

**width of
non-hydrated
sodium ion**

**width of
non-hydrated
potassium ion**

Integral membrane proteins can form transmembrane channels for the passage of charged molecules. Selectivity can be conferred by the size of the channel, the nature of the polar groups lining the interior and also by peripheral membrane proteins at the channel entrance. In the sodium-selective channel, the polar oxygen atoms lining the channel replace the hydration shell of the sodium and allow passage. The larger potassium ion is impermeant. In the potassium-selective channel, the oxygens are arranged close enough to replace the hydration shell of the potassium ion but too far away to dehydrate the sodium ion. It has to stay hydrated and is then too large to fit through the channel. (The sizes are not to scale.)

Several mechanisms have evolved to facilitate the transfer of charged molecules across biological membranes.

- If the ion is a weak acid or base, association with a proton (weak acid) or dissociation of a proton (weak base) can create an uncharged molecule which can diffuse across the membrane (Figure 7.3).

- Lipid-soluble carrier molecules (*ionophores*) can shield the charge sufficiently to allow the complex to diffuse across the membrane (see Table 7.3). The ionophore complex can be electroneutral (for example nigericin) or charged (for example valinomycin, Figure 7.1).

- A transmembrane complex of protein can create an aqueous pore through the membrane. The size of the pore and the nature of the polar groups lining the interior can provide selectivity for particular ions (Figure 7.2).

## Figure 7.3
**Transport of weak acids and bases across membranes**

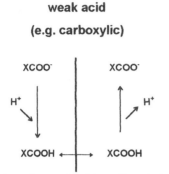

| **weak acid** | **weak base** |
|---|---|
| **(e.g. carboxylic)** | **(e.g. ammonia)** |

The association of a weak acid, or dissociation of a weak base, with a proton allows the uncharged form of the molecule to diffuse through the membrane down its concentration gradient. Proton association or dissociation on the other side of the membrane results in the original ionic form of the molecule. In both cases, a proton gradient is created by the transport. Eventually the pH on each side of the membrane will change sufficiently to inhibit further movement by affecting the association/disassociation reactions.

## 7.3 Osmotic pressure

With some exceptions, biological membranes are relatively permeable to water because of the small size of its molecule. A lowering of the effective concentration (activity) of water is caused by the presence of solutes. Water, like other molecules, will diffuse into regions where it is less concentrated. Most solutes in the cell are not able to diffuse across cell membranes and therefore water concentration differences will be created. The water will then diffuse across the membrane and cause the compartment to swell. The pressure required to prevent the water movement is called the *osmotic pressure* and, unless this is equalised, the compartment will burst. Osmotic pressure is a colligative property, it depends on the *number* of molecules in the system. Thus 0.5 M sucrose has an osmotic concentration of 0.5 Osm whereas 0.5 M potassium chloride has an osmotic concentration of 1 Osm. For dilute solutions, the osmotic pressure ($\Pi$) can be given as

$$\Pi = cRT$$

where $c$ is the osmotic concentration, $R$ is the gas constant (usually expressed for osmotic calculations as 0.082 litre-atm mol$^{-1}$ K$^{-1}$) and $T$ is the temperature (K). In cells, the osmotic concentration of impermeant molecules is around

**Table 7.3** Some ionophores and their mediated transport properties

| Ionophore | Specificity | Mechanism | |
|---|---|---|---|
| valinomycin | $K^+$, $NH_4^+$ $(H^+)^a$ | uniport, electrogenic (mobile carrier) | |
| nigericin | $K^+/H^+$ | antiport, electroneutral (mobile carrier) | $K^+$ / $H^+$ |
| monensin | $Na^+>K^+/H^+$ | antiport, electroneutral (mobile carrier) | $Na^+$ / $H^+$ |
| A23187 | $Ca^{2+}/2H^+$ | antiport, electroneutral (mobile carrier) | $Ca^{2+}$ / $2H^+$ |
| gramicidin | $H^+>K^+$, $Na^+$ | uniport, electrogenic (channel former) | $Na^+$ $H^+$ $K^+$ |

[a]Valinomycin has high specificity for $H^+$, around the same as for $K^+$. However, under normal physiological conditions at pH 7, where the concentration of $H^+$ is $10^{-7}$ M and $K^+$ is in the millimolar range, $K^+$ is effectively the only ion transported.

0.3 Osm, equivalent to 8 atmospheres of pressure. Biological membranes are not strong enough to withstand this pressure difference and would rupture without any compensating force. Organisms have to pay an energy price to prevent this happening. In animal cells, the active extrusion of sodium ions keeps the surrounding fluid effectively *isotonic* (of equal osmotic pressure) with the fluid within the cell. The biosynthesis of cell walls in plants, fungi and bacteria helps support the plasma membrane and enable the cells to withstand hypotonic environments without bursting. Turgor pressure has an additional advantage for plant stems and leaves in providing mechanical support. In many protozoa, water is actively extruded using special contractile vacuoles.

## 7.4 Active transport

### 7.4.1 *Competing solutes*

Consider a non-specific sugar transport protein that will transport sugars across a cell membrane. Initially, let there be equal concentrations of a sugar, xylose say, across the membrane. Then let us add another sugar, glucose say, to a higher concentration on one side of the membrane. What will happen?

|       outside       |       inside       |
|:-------------------:|:------------------:|
| $C_o$ (glucose)     |                    |
| $C_o$ (xylose)      | $C_i$ (xylose)     |

Xylose and glucose will compete for the transporter on the outside face of the membrane and, since glucose is at the higher concentration, it will preferentially bind to the transporter and be transported down its concentration gradient. However, once on the inside face of the membrane the transporter will be exposed to higher concentrations of xylose compared to glucose and will therefore exchange glucose for xylose. The xylose will be transported back across the membrane where the cycle will start again. The transport of glucose down its concentration gradient will drive a counterflow of xylose in the opposite direction. A gradient of xylose will be created by active transport (against its concentration gradient) by coupling the process to the free energy change of the gradient of glucose (down its concentration gradient). Net flux will stop when the two gradients balance. This can be quantified using Equation (7.1).

The same principle applies to charged molecules with the additional involvement of the energy associated with the membrane potential (Equation (7.2)). Membrane gradients, whether of charged or uncharged molecules, can be used to drive the active transport of other molecules provided there is a specific transporter to catalyse the transport. The transporter can transfer both solutes in the same direction (*symport*) or in the opposite direction (*antiport*). One physiological example of coupled transport is the co-transport of sodium and glucose in the epithelial cells of the small intestine (Figure 7.4). Glucose is actively concentrated in the cell by sodium-driven symport at the apical surface of the cell. It

*Figure 7.4*
**The intestinal epithelial cell**

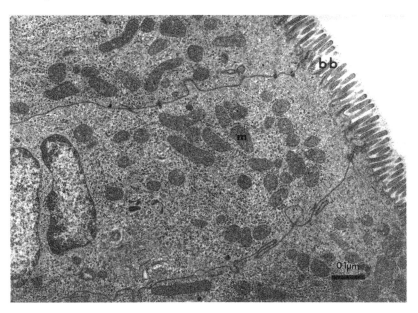

An electron micrograph of a mammalian epithelial cell showing the nucleus, brush border (bb) and numerous mitochondria (m). Various transport systems catalyse the movement of solutes from the lumen to the cytosol across the brush border membrane. (Courtesy John Pacy, Electron Microscope Unit, King's College London.)

then diffuses into the bloodstream via uniport glucose transporters on the basolateral surface of the cell. The sodium gradient is created in the first place by an active $Na^+/K^+$-ATPase in the plasma membrane.

### 7.4.2  *Coupling chemical reactions to transport*

A chemical reaction can be coupled to drive the transport of molecules across a membrane provided a suitable common intermediate is present. A good example is the *phosphotransferase system* for the transport of hexose sugars in many bacteria (Figure 7.5). The ultimate driving force for the sugar transport is the free energy associated with the hydrolysis of phosphoenolpyruvate (see Table 3.2). The mechanism of coupling involves the sequential transfer of a phosphate group from phosphoenolpyruvate to the sugar via a phosphorylated enzyme intermediate. The phosphotransferase systems found in different bacteria are not

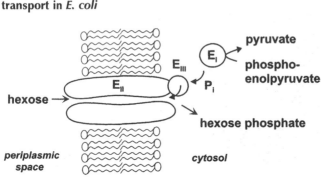

*Figure 7.5*
**The phosphotransferase system for sugar transport in *E. coli***

hexose →

periplasmic space

cytosol

pyruvate

phospho-
enolpyruvate

hexose phosphate

The sugar is phosphorylated as it passes through the sugar-specific transmembrane protein ($E_{II}$). The phosphate group is transferred to a component of the membrane protein compex ($E_{III}$) by a separate cytoplasmic enzyme ($E_I$) which catalyses the initial hydrolysis of phosphoenolpyruvate. Phosphotransferase systems differ slightly in composition.

identical but all involve a transmembrane protein which catalyses the final transfer of phosphate to the sugar. Effectively, the sugar is always diffusing down its concentration gradient since the concentration of the nonphosphorylated sugar in the cytoplasm is extremely low.

A superfamily of bacterial ATPases catalyse the transport of a variety of substances across the bacterial plasma membrane. Transported substances include sugars, amino acids and inorganic ions. The systems all share some common features and are named the *ABC family of transporters* (so called because they all contain a highly conserved *A*TP-*B*inding *C*assette, or motif, in their structure). Members of the ABC transporter family are also now being found in eukaryotes (Table 7.4). One important member is the *multidrug resistance (MDR) protein*, an inducible system that pumps a variety of drugs out of cells, making the cells drug-resistant. In animal cells, the transporters comprise polypeptide chains of two structurally similar halves. Each half has six transmembrane helical segments and a cytoplasmic hydrophilic ATP-binding domain. It is not known whether both ATP domains are functional.

A $Na^+/K^+$-ATPase is found in the plasma membrane of nearly all animal cells. Again, the detailed coupling mechanism between the ATP hydrolysis reaction and the ion translocation is not known. It is known that the pump acts as an electrogenic antiporter, exchanging $3Na^+$ ions for $2K^+$ ions against the electrochemical gradient of both ions for every ATP molecule hydrolysed (Figure 7.6). During the catalytic cycle, phosphorylation of an aspartic acid residue of the ATPase takes place on the cytoplasmic side (the involvement of a phosphorylated intermediate classifies this transporter

**Table 7.4**  The ABC family of transporters

The ABC transport families in bacteria and eukaryotes are similar in amino acid sequence and are thought to have evolved from some common ancestral protein. They pump various compounds out of the cell at the expense of ATP hydrolysis.

|  | Examples |
|---|---|
| **Prokaryote** | chloroquine translocator (confers chloroquine resistance on *Plasmodia falciparum* the virulent malarial parasite)<br>oxyanion translocator (confers insensitivity to arsenical and antimonial drugs on a variety of species) |
| **Eukaryote** | multidrug translocator in mammalian cells (confers multidrug resistance (MDR) over a wide spectrum of drugs)<br>cystic fibrosis transmembrane conductance (CFTC) regulator<br>pheromone exporter in yeast<br>multidrug translocator in parasitic protozoa |

*Figure 7.6*
**The Na$^+$/K$^+$-ATPase**

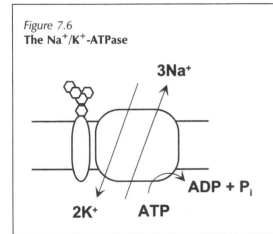

**3Na$^+$**

**2K$^+$**     **ATP**     **ADP + P$_i$**

The basic unit of the Na$^+$/K$^+$-ATPase comprises two subunits, a glycosylated $\beta$-subunit and a larger $\alpha$-subunit. In the plasma membrane the complex exists as a dimer ($\alpha_2, \beta_2$). Hydrolysis of ATP on the $\alpha$-subunit is coupled to the transport of three sodium ions out of the cell and two potassium ions into the cytoplasm, against their concentration gradients. Because the coupling mechanism involves a phosphorylated intermediate, the pump is classed as a P-type pump.

as a *P-type ATPase*). Conformational changes in the protein, induced by phosphorylation, transfer bound sodium ions across the membrane. Binding of potassium ions, followed by dephosphorylation, transfers the potassium inwards and returns the protein to its original conformation. Two conformational states of the ATPase have been identified, one stabilised by sodium ions and the other by potassium ions. The free energy of ATP hydrolysis is coupled to ion transport by changes in protein conformation and binding affinities for the two ions. A similar mechanism of coupling is present in the P-type Ca$^{2+}$-ATPases of the plasma membrane and sarcoplasmic reticulum of animal cells.

Animal cells attempt to maintain lower concentrations of sodium ions in the cytosol than in the surrounding fluid and the opposite concentration ratio for potassium ions (Table 7.5). Using Equation (7.2) and the concentration values in Table 7.5, we can calculate the energy required. For one sodium ion transported against a membrane potential of 60 mV, we have

$$\Delta G = +RT \ln \Delta C_{Na^+} + F \Delta \psi$$
$$= 8.314 \times 310 \times \ln 14 + 96\,485 \times 0.06$$
$$= +12\,591 \, \text{J} \, \text{mol}^{-1}$$

Three moles of sodium ions transported would therefore require 37.8 kJ.

For potassium, although transport is against its concentration gradient, the membrane potential favours the process:

$$\Delta G = +RT \ln \Delta C_{K^+} - F \Delta \psi$$
$$= 8.314 \times 310 \times \ln 30 - 96\,485 \times 0.06$$
$$= +2977 \, \text{J} \, \text{mol}^{-1}$$

For 2 moles of potassium the energy required would be 5.9 kJ. The total energy needed to transport 3 moles of sodium and 2 moles of potassium will therefore be 43.7 kJ. Under cytosolic conditions, the free energy change for ATP hydrolysis is around 45 kJ mol$^{-1}$ and therefore sufficient free energy is available provided tight coupling is observed. Different values for the plasma membrane potential will alter the figures slightly.

**Table 7.5**   Sodium and potassium concentrations inside and outside a typical animal cell

The cell maintains these ratios by active transport of sodium and potassium. A resting membrane potential of around 60 mV (+ve outside) is usually present. Electroneutrality in the aqueous phase is maintained by an equal number of anions, mainly chloride.

| Ion | Inside cell (mM) | Outside cell (mM) | Ratio (in/out) |
| --- | --- | --- | --- |
| Na$^+$ | 10 | 140 | 0.07 |
| K$^+$ | 150 | 5 | 30 |

## 7.5 Coupling redox reactions to ion transport

Several mechanisms have been discovered for coupling the redox reactions of electron transport chains to ion transport. Protons are nearly always the ions involved, although some sodium coupling systems have been identified. The main mechanisms are illustrated in Figure 7.7.

*Figure 7.7*
**Mechanisms for proton translocation by membrane electron transport chains**

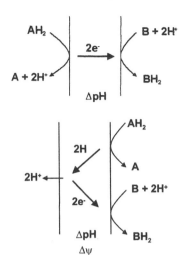

(a) The Lundegårdh mechanism involves the oxidation of a protonated substrate on one side of the membrane and the reduction of an oxidant at the other side with an accompanying proton uptake. The uptake and release of protons generates a pH difference across the membrane.

(b) The Mitchell loop involves the action of alternating proton and electron carriers across the membrane. The release of protons on one side and the return of electrons to the other side generates $\Delta\psi$ as well as $\Delta$pH.

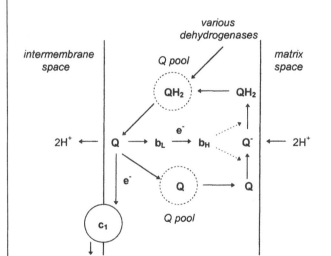

(c) In the Q-cycle, one electron essentially cycles from one side of the membrane to the other during an alternate oxidation and reduction of coenzyme Q on each side of the membrane. On the matrix side it requires two electrons from two turns of the cycle (dotted arrows) to produce one quinol. This effectively causes net oxidation of the quinol pool.

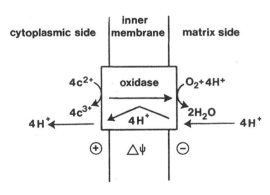

(d) The proton pump, illustrated here for cytochrome oxidase, couples the free energy of the redox reaction to the translocation of protons against their electrochemical gradient. In cytochrome oxidase the oxygen reduction reaction also provides half of a Lundegårdh mechanism.

The simple *Lundegårdh mechanism* involves the oxidation of a protonated substrate on one side of the membrane and the reduction with accompanying proton association of an oxidant at the other side. The number of protons translocated across the membrane (in this case an apparent translocation) for every electron involved in the redox reaction is termed the $H^+/e^-$ ratio and in the example shown in Figure 7.7a has a value of 1 (note the ratio is the same for a one-electron redox reaction ($1H^+/1e^-$) as it is for a two-electron reaction ($2H^+/2e^-$)) The cytochrome *bd*-ubiquinol oxidases of *E.coli* and *Azotobacter vinelandii* function in this way, with the two-electron oxidation of ubiquinol releasing two protons on the periplasmic side of the membrane. Two protons are then taken up from the cytoplasmic side for every oxygen atom reduced. The value need not be a whole number. Oxidation of cytochrome $c^{2+}$ by cytochrome *c* oxidase on the cytoplasmic surface of the mitochondrial inner membrane is not accompanied by proton dissociation from cytochrome *c*. Yet two protons are taken up from the matrix compartment for each oxygen atom reduced. This part-reaction in the catalytic cycle of cytochrome *c* oxidase has an $H^+/e^-$ ratio of 0.5.

The second mechanism, illustrated in Figure 7.7b, is the classical *Mitchell loop* (named after Peter Mitchell, Nobel Prize for chemistry 1978). Oxidation of a reductant on one side of the membrane is accompanied by translocation of protons and electrons across the membrane. This requires the involvement of protonated redox cofactors such as flavin or quinol (see Figure 4.1). Once on the other side, the protons are released and the electrons return, this time carried on electron carriers, for example iron–sulphur complexes or *b*-type cytochromes. The $H^+/e^-$ ratio is unity. The dissimilatory nitrate reductase of *Paracoccus denitrificans* (see Section 6.2) is considered a good example of this type of mechanism, the proton-carrying arm comprising ubiquinol and the electron-carrying arm comprising two *b*-type haems.

The third mechanism, shown in Figure 7.7c, is the *Q-cycle*, probably the best-characterised of all the redox-linked proton translocation systems. The development of the theory came about because of some unexplained experimental results on complex III of the mitochondrial electron transport chain. Essentially it was found that electron transport from ubiquinol to cytochrome *c* was accompanied by proton translocation with an $H^+/e^-$ stoichiometry of 2. A classical loop mechanism involving ubiquinol might be expected to give an $H^+/e^-$ stoichiometry of 1. Another strange experimental finding was that adding a pulse of an oxidant, such as ferricyanide, to the complex under certain conditions induced a transient *reduction* of a *b*-type cytochrome. In a linear scheme, it might be expected that any redox intermediates would become more oxidised. Mitchell realised that these observations could be reconciled by a cyclic scheme in which one electron essentially cycles from one side of the membrane to the other. In one direction it is accompanied by a proton and in the other it travels alone via the two *b*-type cytochromes (Figure 7.7c). The final scheme, like most discoveries in science (in retrospect) is elegant and simple. Ubiquinol from the activity of the various dehydrogenases diffuses to a Q-binding site on complex III at the cytoplasmic side of the inner mitochondrial membrane (termed the p-face, for positive potential). Here it is oxidised and releases two protons to the intermembrane space. However only one electron transfers down the chain to cytochrome *c* (via the Rieseke protein and cytochrome $c_1$). One remains as a semiquinone at the Q-binding site, eventually to transfer back across the membrane via two *b*-type cytochromes. Thus two protons are released on the p-side for only one electron passing to cytochrome *c*. At the matrix side (or n-face for negative) the electron that has 'recycled' is transferred to an oxidised quinone to form another semiquinone (tightly bound at a second quinone-binding site). This awaits a further electron from the *b*-cytochrome pathway before being fully reduced and released as ubiquinol. The protons required are taken from the matrix compartment. Overall, two protons are released for every one electron transferred from ubiquinol to cytochrome *c*, a stoichiometry of $2H^+/e^-$. The effect of increasing the oxidation of cytochrome *c* under steady-state conditions will be to increase the oxidation of ubiquinol at the p-face. In turn this will increase the *reduction* level of the *b*-cytochromes, as observed.

Although the cyclic scheme is simple as described here, there are some puzzles in the detail. The first is thermodynamic. On first sight, the transfer of an electron from the p-face to the n-face involves transferring it from a more positive redox couple ($E^{\circ\prime}$ quinol/semiquinone $\approx +0.28\,\mathrm{V}$) to a much more negative potential ($E^{\circ\prime}$ quinone/semiquinone $\approx -0.16\,\mathrm{V}$). The tight binding of the semiquinones makes this possible. A 10-fold difference in binding of the semiquinone relative to its fully reduced or fully oxidised form will alter the redox potential by 0.06 V (see Equation (4.9) and Figure 3.3). The redox potential of the quinol/semiquinone on the p-face will be made more negative and that for the semiquinone/quinone more positive. A several hundred-fold binding difference will bring the potentials of the two couples together. Of course, the whole cycle is driven by the redox difference between ubiquinol and cytochrome $c$ (a difference of around 0.3 V at steady state).

The second apparent puzzle is that to reduce the semiquinone on the n-face requires a second electron from the cytochrome $b$ pathway. However, getting the second electron requires the oxidation of another ubiquinol on the p-face. In effect we release four protons at the n-face for every two protons taken up at the p-face. Where do these protons come from? In fact they simply come from the *net* oxidation of one ubiquinol at the p-face. During the transfer of two electrons to cytochrome $c$, one ubiquinone is reduced at the n-face but two ubiquinols are oxidised at the p-face. An analogous situation is seen for the cytochrome oxidase reaction. There is a *net* consumption of two protons at the n-face for every water molecule formed.

The fourth mechanism discovered linking electron transfer to proton translocation is the *proton pump* (Figure 7.7d). It was found experimentally by Wikström that cytochrome oxidase translocated protons across the inner mitochondrial membrane during electron transfer from cytochrome $c$ to oxygen. For every proton taken from the matrix compartment (n-face) for oxygen reduction, another proton is translocated from the n-face to the p-face. The mechanism of transfer is still a puzzle. There seems no obvious carrier of protons and electrons in the enzyme complex. The copper ions and haems $a$ and $a_3$ are all electron carriers. Proton carriers such as flavin or quinol are not found and in any case would have too negative a redox potential for this region of the electron transport chain. Mitchell and coworkers suggested that oxygen intermediates might fulfil this function. Several O-loops and O-cycles were suggested, examples being the transfer of hydrogen peroxide in one direction and molecular oxygen or water in the other. Later, Mitchell refined an O-carrier mechanism to alternating hydroxide and water binding on $Cu_B$ as it cycles between its oxidised and reduced forms.

### 7.5.1  *The energetics of proton translocation*

The energy required to translocate protons across the membrane of an electron transport chain can be quantified by Equation (7.2), remembering that the direction of proton translocation is to the p-side of the membrane where the potential is positive (unfavourable) and where the proton concentration is higher (also unfavourable).

$$\Delta G = +RT \ln \frac{[H^+]_p}{[H^+]_n} + F \Delta \psi$$

Since $-\log[H^+]$ is a measure of pH, we can write

$$\Delta G = -2.3RT \, \Delta pH + F \Delta \psi \tag{7.9}$$

The quantity $\Delta G/F$ has been defined by Mitchell as the *protonmotive force* ($\Delta p$) which has units of voltage. Substituting for the constants $R$ and $T$ (at 25°C) and expressing the final equation in terms of mV instead of V, we have

$$\Delta G/F \equiv \Delta p = \Delta \psi - 59 \, \Delta pH \tag{7.10}$$

Thus the protonmotive force is made up of an electrical component, $\Delta \psi$, and a concentration gradient component, $59 \, \Delta pH$ (expressed in the same units as $\Delta \psi$).

In respiring mitochondria, where the respiratory rate is limited by the availability of ADP (state 4), the value of the protonmotive force is around 220 mV, typically made up from a membrane potential of 190 mV and a pH gradient of 0.5. The energy required to translocate protons against this protonmotive force can be calculated for each segment of the respiratory chain (Figure 7.8). The stoichiometry for complex I is not known, but present experimental estimates suggest an $H^+/e^-$ ratio of 2. Thus four protons, together with their four positive charges, are translocated for every pair of electrons passing through the complex, a total energy requirement of $69.4 \, \text{kJ} \, \text{mol}^{-1}$ $NAD_{red}$ oxidised. The energy requirement for complex III is less, since only two charges are translocated although four protons appear at the p-face. On the other hand, complex IV (cytochrome oxidase) translocates two protons and four charges across the membrane. Two positive charges are carried on the translocated protons and two negative charges are carried in the opposite direction by the electron pair to the oxygen reduction site. The energy for the generation of the protonmotive force comes from the redox reactions of the three individual complexes and, as shown in Chapter 5, is sufficient to accommodate the proposed schemes shown in Figure 7.7.

### 7.5.2 *Sodium-motive systems*

The expression for the protonmotive force is only specific to protons with regard to $\Delta pH$. $\Delta \psi$ could be created or used by any charged molecule if a coupling mechanism were available. In many bacteria, the circulation of sodium ions is maintained across the cytoplasmic membrane by coupling it to proton movement by $Na^+/H^+$ antiporters. The toxic

*Figure 7.8*

**The energetics of proton translocation by
the respiratory chain of mitochondria**

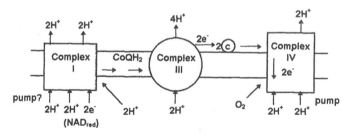

|  | Complex I | Complex III | Complex IV |
|---|---|---|---|
| $\Delta E'$ for span (mV) | 290 | 310 | 380 |
| $\Delta G$ available per electron pair (kJ) | -56.0 | -59.8 | -73.3 |
| Protons transferred per electron pair | 4 | 4 | 2 |
| $\Delta G$ required * (kJ) | +7.7 | +7.7 | +3.9 |
| Charge transferred per electron pair | 4 | 2 | 4 |
| $\Delta G$ required* (kJ) | +61.7 | +30.9 | +61.7 |
| Total $\Delta G$ required for proton translocation per pair of electrons (kJ) | +69.4 | +38.6 | +65.6 |

* Assuming a protonmotive force in State 3 of 180 mV ($\Delta\psi \equiv 160$ mV; $\Delta$pH $\equiv 20$ mV)

Different complexes of the respiratory chain translocate different numbers of protons and charge. For example, complex III translocates four protons for every two electrons passing through the complex but only two charges. This is because electron entry into the complex is on the matrix side but electron delivery is to cytochrome $c$ on the outside of the membrane. Hence the complex only 'feels' $\Delta\psi$ to half the extent it does $\Delta$pH. Complex IV, cytochrome oxidase, is exactly the opposite. There is an apparent problem with the energetics of complex I if, as assumed, it is associated with a proton pump. The available energy from the redox reaction ($-56$ kJ) is less than that required to move four electrogenic protons from the matrix to the outside ($+69.4$ kJ). Of course the protonmotive force measurements may be in error, and also the assumed redox potentials for the NAD and CoQ couples may be different. Thermodynamics should not take precedence over experiment. Nevertheless, the finding should give pause for thought.

effects of sodium ions are avoided by keeping the internal sodium concentration low, even in the high-salt environments experienced by halophilic bacteria. An $Na^+/H^+$ antiporter couples the protonmotive gradient generated by the respiratory chain to the active extrusion of sodium ions. In eukaryotic microorganisms and in plants, the $Na^+/H^+$ antiporter is involved in the regulation of cytoplasmic pH. Enhanced sodium ion influx (hence proton efflux) has been measured after artificially acidifying the cytoplasm of algal cells. The antiporter appears to be electrogenic, with a possible exchange of two protons for every sodium ion.

Primary sodium-motive systems have also been found. A sodium electrochemical gradient can be generated in some fermenting anaerobes by the action of $Na^+$-transporting decarboxylases. The free-energy from the decarboxylation of methylmalonyl-CoA is coupled to sodium ion extrusion in *Propionigenium modestum*. The electrochemical sodium ion gradient generated can then be used to drive ATP synthesis by a sodium-dependent ATP synthase (see Section 7.6). A respiratory-driven primary sodium pump has been discovered in the aerobe *Vitreoscilla*. Sodium ions replace protons for the generation of $\Delta\psi$. So far, sodium seems to be the only ion to replace protons for the generation of primary electrochemical gradients and the coupling mechanisms at the molecular level are largely unknown.

## 7.6 Chemiosmotic coupling of proton gradients to solute transport

The protonmotive force generated by respiration is widely used to drive the movement of solutes across cellular membranes. $\Delta\psi$ and $\Delta pH$ can have different roles in the transport process. For example, $\Delta\psi$ drives the movement of mitochondrial ATP into the cytoplasm by an electrogenic exchange with cytoplasmic ADP (Figure 7.9). The difference in negative charge between $ATP^{4-}$ and $ADP^{3-}$ means that the exchange favours the movement of ATP to the cytoplasm (+ve) by the *adenine nucleotide translocator*. The translocator catalyses an obligatory 1:1 exchange of the nucleotides. When the nucleotide binding site is exposed to the cytoplasmic side of the mitochondrial inner membrane, binding can be inhibited by *atractyloside*. When binding nucleotides from the matrix compartment, the translocator is inhibited by *bongkrekic acid*. The pH component of the protonmotive force drives the electroneutral transport of inorganic phosphate into mitochondria by a symport mechanism with $H^+$

*Figure 7.9*
**The transport of adenine nucleotides and inorganic phosphate across the mitochondrial inner membrane**

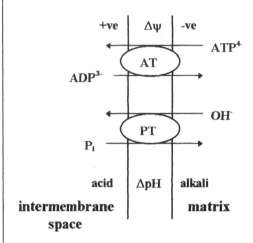

ATP export from the mitochondrial matrix is catalysed by the adenine nucleotide transporter (AT) in exchange for ADP. The antiport is electrogenic and driven by the outside-positive potential of the membrane. Phosphate transport by the phosphate transporter (PT) is electroneutral, in exchange for $OH^-$, and is driven by the $\Delta pH$ across the membrane.

(Figure 7.9). Thus the full component of the protonmotive force, $\Delta\psi - 59\,\Delta pH$, drives ATP from where it is primarily synthesised in the mitochondrion into the cytoplasm where it is mainly used. The ADP and phosphate generated in the cytoplasm can then be transported back into the mitochondrion for further ATP synthesis. There are a variety of other metabolite carriers in the mitochondrial inner membrane, all driven by one or both components of the protonmotive force (Table 7.6).

Bacteria contain a number of proton symport systems where the co-transport of a sugar with a proton, down the proton electrochemical gradient, drives the sugar import. The best-characterised of these is the *lactose (lac) permease* transporter of *E. coli*. Since lactose is uncharged, its co-transport with a proton is electrogenic and therefore $\Delta\psi$- and pH-dependent. A protonmotive force of around 200 mV would support the generation of a lactose gradient of several thousand-fold (see Table 7.1).

## 7.7 Chemiosmotic coupling of proton gradients to ATP synthesis

The coupling of a proton gradient to the phosphorylation of ADP requires a protonmotive ATP synthase. Such an enzyme is found associated with the respiratory chain membranes of

**Table 7.6** Mitochondrial inner-membrane transporters

| Carrier | Movement | | Dependence |
|---|---|---|---|
| Adenine nucleotide | $ADP^{3-}$ →  ← $ATP^{4-}$ | | $\Delta\psi$ |
| Phosphate | $Pi^-$ → <br> $H^+$ → <br> (or $OH^-$ antiport) | | $\Delta pH$ |
| Dicarboxylate | $malate^{2-}$ →  ← $Pi^{2-}$ | | $\Delta pH$ |
| Tricarboxylate | $citrate^{3-}$ $+ H^+$ →  ← $malate^{2-}$ | | $\Delta pH$ |
| Pyruvate | $pyruvate^-$ →  ← $OH^-$ <br> (or $H^+$ symport) | | $\Delta pH$ |

mitochondria, chloroplasts and bacteria. Its structure is now well characterised (Figure 7.10). A transmembrane segment $F_0$ is attached to an extramembrane portion $F_1$ which contains three catalytic sites for ATP synthesis. The driving force for the phosphorylation reactions is supplied by the discharge of a proton gradient through $F_0$. The $F_1$ portion of the enzyme can be detached from the membrane by mild solvent treatment but then can only act as an ATPase. The coupling of the protonmotive force to ATP synthesis requires both $F_1$ and $F_0$.

The crystal stucture of $F_1$ has been determined and comprises a globular domain of six subunits, $3\alpha$ and $3\beta$, plus three other subunits $\gamma$, $\delta$ and $\varepsilon$. The $\gamma$ subunit provides a central helical structure about which the $\alpha_3\beta_3$ domain can rotate (Figure 7.10). The solution of the crystal structure confirms an earlier suggestion by Boyer that catalysis depends on changes in binding energy in each of the nucleotide-

*Figure 7.10*

**The structure of the mitochondrial ATP synthase**

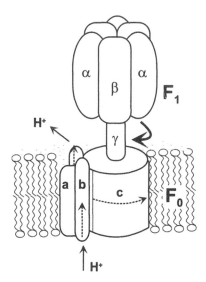

The $F_1$ portion of the enzyme is attached to the transmembrane portion $F_0$ by the $\gamma$-subunit which provides a central helical structure about which the $\alpha_3\beta_3$ domain can rotate. Each 120° rotation induces conformational changes in the $\alpha$ and $\beta$ subunits and generates one ATP from bound ADP and phosphate. The rotation is driven by the movement of protons through $F_0$ via a ring of c-subunits, 9–12 in number depending on the source of the enzyme.

binding sites. The $\beta$ chains provide the catalytic unit for the enzyme although the $\alpha$ chains contribute to the catalytic interface. Each catalytic site can exist in three conformations, each with a different binding affinity for nucleotides and phosphate. As mentioned in Chapter 3, the free energy for a reaction can be altered substantially by a differential binding of reactants and products to a solid phase (see Figure 3.3). In the case of the $F_0F_1$-ATP synthase, one of the three conformations binds ADP and phosphate with high affinity. Water is excluded and ATP formation is favoured. However, the ATP does not exchange easily with the bulk solution and a conformational change to a low-affinity state is required to release the ATP from the catalytic site. Binding of fresh ADP and phosphate takes place after a further conformational change in the catalytic site to a 'loose' binding state. The three catalytic sites on $F_1$ undergo these conformational changes sequentially, each 120° out of phase with its neighbour. At any instant in time, the three catalytic subunits are in different conformational states.

The coupling mechanism requires the movement of protons through $F_0$ to be linked to a sequential alteration of protein structure in $F_1$. Again, the crystal structure provides evidence for a plausible mechanism. The $\gamma$ subunit can be seen to provide a hydrophobic axis around which the $\alpha_3\beta_3$

domain can rotate. It effectively acts as a smooth bearing for a molecular motor. Each $120°$ rotation transduces the mechanical energy in the $\gamma$ subunit into a conformational alteration of the $\alpha_3\beta_3$ structure. The activity of the bovine ATP synthase is around $400\,s^{-1}$. If three ATP molecules were synthesised for each complete rotation, then this would correspond to around 130 Hz. How is the rotation generated? Unfortunately we do not yet have a crystal structure for the $F_0$ complex, but present evidence suggests the $\gamma$ subunit extends into a ring of transmembrane helices made up from 9–12 copies of a small polypeptide. Other polypeptides are thought to catalyse the entry of protons from the p-side of the membrane into a site where interaction occurs with the ring of helices. Binding of a proton to one of the helices, probably to an aspartate residue, causes a rotation of the ring structure by changes in electrostatic forces. This brings the next helix into alignment for proton binding. Eventually the proton is discharged to the n-side of the membrane. Like all motors, the assembly needs a stator to prevent equal and opposite rotation of its different parts. Other protein subunits are thought to fulfil this role.

The $F_1$ and $F_0$ complexes appear to have evolved independently. The $F_1$ complex shows structural similarities to DNA helicases. An $F_0$-like structure is present in the bacterial flagellum motor, which is also driven by a protonmotive force. Sodium-motive ATP synthases have been identified. The specificity of the synthases for sodium or protons seems to reside in the $F_0$ segment. A hybrid synthase comprising the $F_1$ complex from the protonmotive synthase of *E. coli* and the $F_0$ complex of the sodium-motive synthase from *P. modestum* turns out to be sodium motive in function.

### 7.7.1 *P/O ratios*

The P/O ratio (see Box 5.1) for ATP production coupled to electron transport by an electrochemical gradient of protons will simply depend on the number of protons and the charge translocated by each system. If the $H^+/e^-$ ratio for the respiratory chain is $n$ and the $H^+/ATP$ ratio for the synthase is $m$, then the P/O ratio will be $n/m$ (Figure 7.11). Experiment seems to indicate that the $H^+/ATP$ ratio for the mitochondrial ATP synthase is 3. If 10 protons, together with 10 charges, are translocated during two-electron transfer from $NAD_{red}$ to oxygen (Figure 7.8), then a P/O ratio of 3.3 might be expected. Electron transfer from succinate to oxygen ($n = 6$) would theoretically generate a P/O ratio of 2.0. Experimental measurements (see Chapter 5) indicate values

*Figure 7.11*

**The proton circuit linking electron transfer to ATP synthesis**

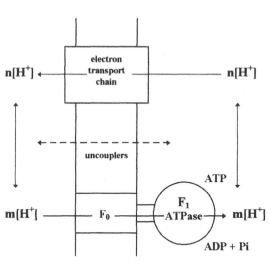

Electron transport by respiratory chains is coupled to ATP synthesis by an electrochemical gradient of protons. Oxidation of substrate (for example, succinate) results in the electrogenic transfer of $n$ protons across the respiratory membrane. ATP synthesis is driven by the favourable transfer of $m$ protons back across the membrane. The ratio $n/m$ gives a measure of the number of ATP molecules formed per reductant oxidised. Uncouplers are compounds that discharge the proton gradient by making the membrane permeable to protons.

of 2.6, 1.6 and 1.2 for two-electron transfer from $NAD_{red}$, succinate and cytochrome $c^{2+}$, respectively. Uncertainties in $H^+/e^-$ and $H^+/ATP$ stoichiometries will affect the theoretical values, but an additional factor is that mitochondria have to export ATP and import ADP and phosphate during coupled ATP synthesis. As shown in Figure 7.9, an additional electrogenic movement of one proton is required for each ATP synthesised. The theoretical values for $NAD_{red}$ and succinate oxidation are therefore lowered to 2.5 and 1.5, respectively. The chloroplast ATP synthase is thought to have a higher $H^+/ATP$ ratio than the mitochondrial enzyme, closer to 4 than 3. which may reflect a larger number of helical subunits in the rotating motor of the $F_0$ complex. The P/O ratio for the photosynthetic electron transport chain (see Chapter 8) will be correspondingly lower.

## 7.8 Respiratory control

Proton generators and proton consumers are usually tightly coupled. Electron transport is inhibited by the protonmotive force. At low matrix concentrations of ADP (state 4), the energy for ATP formation is shifted to an unfavourably high level and the ATP synthase is inhibited. The entry of ADP and phosphate into the mitochondrion, from the ATP hydrolysed in the cytoplasm, relieves the inhibition. Sufficient ADP (and phosphate) is now present to lower the free energy required for ATP formation and the protonmotive force can drive ATP synthesis (state 3). The use of the protonmotive force by the synthase (and by the systems importing ADP and phosphate) lowers its value from around 220 mV to around 180 mV and, in turn, relieves the inhibition on electron transport. The effect of *respiratory control* can be seen in isolated mitochondria by the use of an oxygen electrode (Figure 7.12). Respiratory control is of immense importance to the physiology of animals. It is a primary mechanism for balancing the catabolism of food to energy demand. The response of the electron transport chain to a fall in the ATP/ADP ratio accompanies corresponding increases in citric acid cycle activity and in glycolysis, two other pathways stimulated by the changes in adenine nucleotide levels.

*Figure 7.12*
**Measurement of respiratory control in isolated mitochondria**

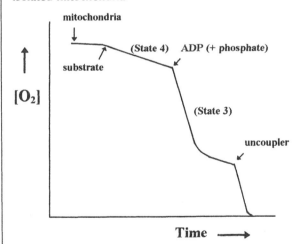

The respiration of mitochondria isolated from a tissue such as liver or heart can be followed using an oxygen electrode. The electrode monitors the oxygen concentration in the mitochondrial suspension. Addition of a substrate such as succinate allows the mitochondria to respire at a rate which is controlled by the protonmotive force generated by electron transfer (state 4). Respiration increases when ADP and phosphate are added (state 3). The import of ADP and phosphate, and the subsequent flux of protons through the $F_0F_1$ATP synthase to generate ATP, lowers the protonmotive force and relieves the inhibition on electron transfer. When most of the ADP is converted to ATP, electron flow is inhibited again by the rise of the protonmotive force. If the amount of ADP added is known, the P/O ratio can be calculated by measuring the oxygen consumed during state 3. The ratio of the respiratory rates of state 3 to state 4 is a measure of the *respiratory control* of the system. Uncouplers act by catalysing the transport of protons down their electrochemical gradient, effectively short circuiting the proton current through the transport systems and the ATP synthase. The free energy stored in the proton gradient is converted to heat.

### 7.8.1 Thermogenesis

If the coupling between electron transfer and ATP synthesis is lost, most of the free energy from the oxidation of food appears as heat. Heat is a necessary by-product of metabolism. The second law of thermodynamics makes this inevitable (see Chapter 1). However, the dissipation of free energy as heat can serve a useful purpose in maintaining body temperature. Extra heat dissipation can be generated as a by-product of physical activity. *Shivering thermogenesis* is a natural mechanism of thermoregulation in mammals. The 'efficiency' of muscular work is low, often around 25%. For every joule of mechanical free energy, three joules are released as heat. Some cold-blooded animals also use metabolic heat production for thermoregulation. Several insects 'warm up' before flight by vibratory movements of their flight muscles. The increased use of ATP activates metabolism and increases heat production. The bumble bee in particular has developed cycles of pre-flight muscle contraction to an extent that the temperature of the thorax can be raised from ambient to over 30°C in a few minutes.

*Non-shivering thermogenesis* produces heat by processes other than muscular work. The main contribution to non-shivering thermogenesis in mammals is the ability of mito-

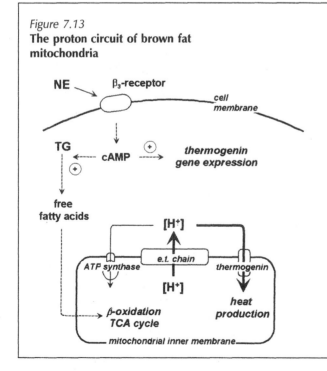

*Figure 7.13*
**The proton circuit of brown fat mitochondria**

Stimulation of the $\beta_3$-adrenergic receptors of the brown fat cell by norepinephrine (NE) leads to stimulation of triglyceride (TG) breakdown into fatty acids. Oxidation of the fatty acids in mitochondria generates reducing equivalents for the respiratory chain (e.t. chain). The electrochemical gradient of proteins across the inner mitochondrial membrane is dissipated through the trans-membrane protein thermogenin, to generate heat. The uncoupling action of thermogenin is directly stimulated by fatty acid binding and inhibited by GTP. Thermogenin production is stimulated by the rise in cAMP.

chondria in *brown fat tissue* to dissipate the protonmotive force by controlled uncoupling. The mitochondria of brown fat tissue (so called because of the large number of mitochondria in the fat cells compared with normal adipocytes) contain a unique transmembrane protein, thermogenin, which acts as a channel for protons. Binding of norepinephrine to $\beta$-adrenoreceptors stimulates triglyceride breakdown into free fatty acids. The fatty acids are used as a respiratory substrate by the mitochondria and also act to open a proton channel in thermogenin. The free energy of the proton gradient generated by the respiratory chain is released as heat, as the protons re-enter the matrix through the uncoupling protein (Figure 7.13). Brown fat tissue is found in all new-born mammals (apart from the domestic pig, for some strange reason). It serves a functional role as the new-born leaves the thermally regulated environment of the womb. It is less prominent in the adult organism but may still play a role in the regulation of energy balance. The lack of an ability to express the gene for thermogenin may be responsible in part for the development of obesity in the adult mammal.

## Summary

- Molecular gradients across biological membranes can be used to store energy. The free energy in a gradient of a charged molecule is far greater than that for an uncharged molecule with the same concentration gradient. Membrane potentials are created by the unequal separation of charge across the membrane. They can enhance or inhibit the transport of other ions depending on the charge and polarity of the potential.

- The free energy available in a gradient of ions or uncharged molecules can be used by specific transporters to catalyse co-transport and counter-transport of other molecules across the membrane.

- The free energy of chemical reaction such as ATP hydrolysis can be coupled to membrane transport provided suitable common intermediates are present. An example is sugar transport into bacteria, which utilises the free energy from the hydrolysis of phosphoenolpyruvate. An example for charged ions is the sodium/potassium ATPase.

- Redox reactions can be coupled to ion transport, mainly of protons but some sodium-motive systems have been found. Four main mechanisms have been proposed. The Lundegårdh mechanism involves the oxidation of a protonated substrate on one side of the membrane with the release of a proton, and the reduction of an oxidant at the other side with an accompanying proton uptake. The Mitchell loop involves the action of alternating proton and electron carriers across the membrane. In the Q-cycle, one electron essentially cycles from one side of the membrane to the other during the alternate oxidation and reduction of coenzyme Q on each side of the membrane. The final mechanism is the proton pump, where the free energy of the redox reaction is coupled to the translocation of protons against their electrochemical gradient.

- Proton gradients can be coupled to the transport of solutes by specific translocators. Examples include the export of ATP from the mitochondrial matrix and the import of ADP and phosphate from the cytoplasm.

- Redox reactions can be coupled to ATP synthesis by an electrochemical gradient of protons. Proton flow through the $F_0F_1$-ATP synthase catalyses the phosphorylation of ADP to ATP. The coupling mechanism requires the movement of the protons to be linked to a sequential alteration of protein structure in the $F_1$ portion of the synthase enzyme.

- Electron transport in mitochondria is inhibited by the proton gradient. Systems that utilise the free energy of the proton gradient relieve the inhibition and oxygen consumption increases. The phenomenon of respiratory control is a primary mechanism for balancing catabolism in the organism with the energy demand. Making the membrane permeable to protons uncouples the redox reactions from energy-utilizing systems and the free energy is dissipated as heat. The mitochondria in brown fat tissue of mammals contain a transmembrane protein, thermogenin, which acts as a channel for protons. Oxidation of fat by these mitochondria generates heat for thermal regulation.

## Selected reading

Cannon, B. and Nedergaard, J., 1985, The biochemistry of an inefficient tissue: brown adipose tissue, *Essays in Biochemistry* **20**, 110–164. (A clear description of the uncoupling system of brown fat mitochondria)

Dimroth, P., 1991, $Na^+$-coupled alternative to $H^+$ coupled primary transport system in bacteria, *Bioessays* **13**, 463–468. (Review of an interesting system for coupling sodium electrochemical gradients to ATP synthesis)

Harris, D. H., 1995, *Bioenergetics at a Glance*, Oxford: Blackwell Science. (A useful summary of molecular bioenergetics)

Heinrich, B. and Esch, H., 1994, Thermoregulation in bees, *Am. Sci.* **82**, 164–171.

Henderson, P.J.F., 1991, Sugar transport proteins, *Curr. Opinion Struct. Biol.* **1**, 590–601.

McNab, R.M., 1990, The genetics, structure and assembly of the bacterial flagellum, *Symp. Soc. Gen. Microbiol.*, **46**, 77–106. (Useful to compare the bacterial flagellum motor with the $F_0$ portion of the $F_0F_1$-ATP synthase)

Mitchell, P., Mitchell, R., Moody, A.J., West, I.C., Baum, H. and Wrigglesworth, J.M., 1985, Chemiosmotic coupling in cytochrome oxidase: possible protonmotive O-loop and O-cycle mechanisms, *FEBS Lett.* **199**, 1–7. (Proposed mechanisms for proton translocation based on oxygen intermediates)

Mitchell, P., 1987, A new redox loop formality involving metal-catalysed hydroxide-ion translocation, *FEBS Lett.* **222**, 235–245. (A proposed mechanism for a proton pump based on an hydroxide-ion intermediate)

Nicholls, D.G. and Ferguson, S.J., 1992, *Bioenergetics 2*, London, San Diego: Academic Press. (An excellent and thorough treatment of the chemiosmotic theory for students)

Postma, P.W., Lengeler, J.W. and Jacobson, G.R., 1993, Phosphoenolpyruvate:carbohydrate phosphotransferase systems of bacteria, *Microbiol. Rev.* **57**, 543–594.

Trayhurn, P. and Nicholls, D.G., eds., 1986, *Brown Adipose Tissue*, London: Edward Arnold.

Wikström, M., 1977, Cytochrome oxidase is a proton pump, *Nature* **266**, 271–273. (The first experimental demonstration of a redox linked proton pump)

### *Structure*

Abrahams, J.P., Leslie, A.G.W., Lutter, R. and Walker, J.E., 1994, Structure at 2.8 Å resolution of $F_1$-ATPase from bovine heart mitochondria, *Nature* **370**, 621–628.

## Study problems

1.  Calculate the average distance moved in 1 s by glycine in the cytoplasm of a cell. (Assume the diffusion coefficient for glycine under these conditions is $100 \times 10^{-11} \, m^2 \, s^{-1}$).

2.  What is the osmotic pressure at 20°C across a semi-permeable membrane with 0.5 M NaCl on one side and pure water on the other?

3.  Calculate the free energy required to pump 2 moles of calcium ion from the cytosol (0.1 $\mu$M) to the extracellular fluid (1 mM) by the $Ca^{2+}$-ATPase, against a membrane potential of 60 mV. What energy would be required if this were an electroneutral event?

4.  What membrane potential (negative inside) will be needed to form a 60-fold concentration gradient of a

sugar inside a cell if the carrier molecule co-transports one sodium ion with every sugar molecule?

5. A suspension of purified mitochondria in 5 ml of a buffered medium containing inorganic phosphate and 2-oxoglutatate was found to consume oxygen at a rate of 0.12 $\mu M$ $min^{-1}$. This increased to 0.6 $\mu M$ $min^{-1}$ following the addition of 100 nmol of ADP. After 5 min the oxygen utilization fell back to its original rate. What is the respiratory control ratio for the mitochondria under these conditions? Calculate the apparent P/O ratio for the oxidation of 2-oxoglutarate. Interpret your answer.

6. During the transition from moderate to intense exercise, oxygen consumption can double and blood lactate can increase more than 10-fold. Discuss the biochemistry of the muscle cell during these changes.

# 8 Energy Capture: Photosynthesis

Over 99% of the free energy in the biosphere arises from the sun. The remaining free energy is obtained from the oxidation of inorganic compounds such as hydrogen and sulphur by specialist microorganisms, the chemolithotrophs (see Chapter 4). *Photosynthesis* is the process in which living organisms, phototrophs, convert the light energy from the sun into the chemical energy of bonding electrons.

## 8.1 Light energy

Electromagnetic radiation has a spectrum spreading from $\gamma$-rays (wavelength <0.001 nm) to radio waves (>1 m). Visible light falls in the middle wavelength range from the violet-blue end of the spectrum at 400 nm to the far red at around 700 nm. The velocity of light ($c$) in a vacuum is constant ($3 \times 10^8\,\mathrm{m\,s^{-1}}$) and is related to frequency ($\nu$) and wavelength ($\lambda$) by the formula

$$c = \nu\lambda$$

Electromagnetic radiation interacts with matter as though it existed in small packets (photons) with discrete energy values (quanta). The quantum of energy ($\varepsilon$) of a *photon* is related to the frequency (and hence wavelength) of the radiation by the expression

$$\varepsilon = h\nu = \frac{hc}{\lambda} \tag{8.1}$$

where $h$ is Planck's constant ($6.62 \times 10^{-34}\,\mathrm{J\,s^{-1}}$). Radiation increases in energy as the wavelength becomes smaller or as the frequency increases. The energy of photons can easily be calculated (see Table 8.1 for the visible light region) and is usually expressed in terms of moles of photons (or *Einsteins*). Another useful way of expressing the energy is in electronvolt units. One electronvolt (1 eV) is the energy acquired by an electron when it moves through a potential of 1 V and is equal to $1.6 \times 10^{-19}\,\mathrm{J}$.

Ozone in the earth's atmosphere shields the surface of the earth from potentially damaging ultraviolet radiation (see Box 6.1). Other components of the atmosphere, notably water vapour and carbon dioxide, trap some of the infrared

**Table 8.1**  Energies of visible light

| Colour | Wavelength (nm) | Energy (kJ Einstein$^{-1}$)[a] | eV/photon |
|---|---|---|---|
| blue | 400 | 299 | 3.10 |
| yellow | 550 | 134 | 2.25 |
| red | 700 | 171 | 1.77 |

[a]An Einstein is Avogadro's number of photons ($6.023 \times 10^{23}$).

(heat radiation) that is re-radiated from the surface of the earth and act to maintain a warmed atmosphere. Without this atmospheric blanket, the surface of the earth would be too cold to sustain life as we know it (down to a mean of $-18°C$ instead of the present mean of $+15°C$). Unfortunately, increases in the concentrations of carbon dioxide and other heat-absorbing molecules from the burning of fossil fuels over the past hundred years are thought to have caused increases in atmospheric warming. The so-called *greenhouse effect* may have unforseen impact on our environment (see Box 8.1).

The standard free energy change for the oxidation of glucose to carbon dioxide and water is $-2870 \, kJ \, mol^{-1}$. To reverse the process and synthesise glucose from six molecules of carbon dioxide would take at least $+2870 \, kJ \, mol^{-1}$. To fix 1 mole of carbon dioxide would therefore require $+478 \, kJ \, mol^{-1}$. If we take red light at 700 nm as the lowest-energy light capable of being used in photosynthesis (although some bacteria can use light up to 900 nm), then at least 2.8 moles of quanta per mole of carbon dioxide would be needed (478/171). In fact three to four times that number is found for most photosynthetic processes.

## 8.2 Mechanisms of light capture

There are three main systems for photon capture based on the type and organisation of the light-absorbing pigments.

- **Bacteriorhodopsin-containing systems**. Only found in Halobacteriaceae such as the aerobic photosynthetic archaebacterium *Halobacterium halobium*. The light energy is transduced directly into the energy of an electrochemical gradient of protons by a light-driven proton pump. The free energy of the gradient is then coupled to other biosynthetic reactions.

- **Bacteriochlorophyll-containing systems**. Found in the green and purple bacteria such as Chlorobiaceae and

*Box 8.1*  **The greenhouse effect**

The greenhouse effect is so called because of the similarity to what happens to the temperature inside a greenhouse when the sun is shining. Light enters and is absorbed by the material in the greenhouse. As this warms up it re-emits longer-wavelength energy (heat radiation) which is trapped by the glass rather than escaping to the outside. Re-emission of heat by the glass throws a certain percentage of the heat back into the greenhouse and increases the temperature.

Different components of the atmosphere absorb different wavelengths of sunlight. Potentially damaging short-wave radiation is absorbed by ozone (see Box 6.1). Most visible light passes through to fall on the surface of the earth with a total energy flux of around $3 \times 10^{16}$ J s$^{-1}$. Infrared radiation approximately doubles this amount. Around half of the solar energy falling on the earth is absorbed by the oceans and land, the rest being reflected back into space. Some of the infrared radiation from the warmed surface of the earth is absorbed by water vapour and carbon dioxide in the atmosphere, and re-radiated back to the surface. A balance is struck between heat gain and heat loss which results in a mean average global temperature of around 15°C.

Water vapour, which can be distinguished from water droplets in cloud and mist, is a very effective 'greenhouse gas' and contributes to more than 80% of the atmospheric greenhouse layer. It has very strong absorption bands in the infrared. Carbon dioxide also absorbs strongly in the infrared but at much longer wavelengths (Table 8.2). There is an absorbance gap between 8000 nm and 12000 nm in which other gases such as methane absorb strongly. The presence of small amounts of a gas can have a disproportionate effect on the atmospheric absorption of radiation if its absorption bands happen to fill a window in the spectrum.

The burning of fossil fuels over the past 100 years has raised atmospheric carbon dioxide levels from around 280 ppm to 350 ppm and the rate of increase is still rising. Similarly, methane concentrations have risen from 0.9 ppm to 1.7 ppm. The consequences for global warming are difficult to predict. However, mean global temperatures appear to have risen around 0.5°C since the late nineteenth century. This may seem insignificant, but a rise in temperature of a few degrees will have a profound effect on climate and agriculture.

**Table 8.2**  The greenhouse gases

| Gas | Atmospheric concentration (ppm) | Main infrared absorption bands (μm) | Relative contribution to greenhouse effect (%) |
|---|---|---|---|
| water vapour | ~3000 | 2.5 –2.9; 5.5–7.1 | >80[a] |
| carbon dioxide | 350 | 4.2; 12–18 | 10–15 |
| methane | 1.7 | 7.6 | 1.0 |
| nitrous oxide | 0.3 | 7.8 | 0.2 |
| ozone | $10^{-1}$–$10^{-3}$ | 9.6 | 0.3 |
| chlorofluorocarbons | $10^{-3}$–$10^{-4}$ | ~9 | 0.3 |

[a]Estimates for water vapour vary between 70% and 90%.

Rhodospirillaceae. These are photosynthetic anaerobes using reductants such as hydrogen sulphide and hydrogen (*photolithotrophs*) or partially reduced organic compounds such as malate or succinate (*photoheterotrophs*) as a source of reducing equivalents.

- **Chlorophyll-containing systems**. Found in green plants, algae and cyanobacteria. These are *photoautotrophs* using water as a source of electrons to reduce carbon dioxide to organic material. Oxygen is released as a waste product.

### 8.2.1 Bacteriorhodopsin-containing systems: Creation of an ion gradient

Halobacteria live in the high salt concentrations (up to 5 M sodium chloride) of shallow salt lakes and usually derive their free energy from the aerobic oxidation of organic compounds. However, when food or oxygen is scarce they switch to photosynthetic activity and generate a protonmotive force by a light-driven proton pump in the plasma membrane. The light-sensitive molecule at the heart of the pump is *retinal* (Figure 8.1), a molecule derived from $\beta$-carotene and also found in the human eye. Retinal is attached to a specific lysine residue (Lys-216) of a transmembrane protein to form bacteriorhodopsin. The relatively small protein (26 kDa) comprises seven transmembrane $\alpha$-helices which cluster with their apolar sides facing the lipid bilayer. Polar amino acid side-chains on each helix face inwards to form a channel for proton conductance (Figure 8.2). The main principles of the proton pumping mechanism are now known (Figure 8.3). Proton uptake and release is caused by transient changes in p$K_a$ values of the Schiff base in

*Figure 8.1*
**All-*trans* and 13-*cis* retinal**

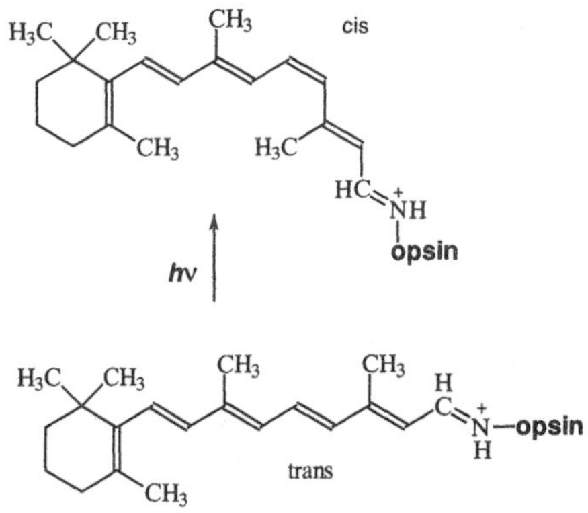

A light-induced *trans* to *cis* isomerisation moves the position of the protonated nitrogen. In bacteriorhodopsin the nitrogen is moved to a more hydrophobic environment. This shifts its p$K_a$ to a lower value and releases the proton to the outside of the membrane.

*Figure 8.2*
**Bacteriorhodopsin**

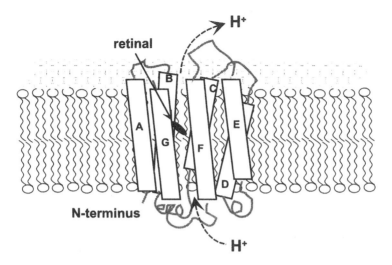

The seven transmembrane helices (A–G) form a proton channel through the membrane. Proton conductance is controlled by light-induced conformational changes in retinal bound to helix G.

*Figure 8.3*
**The photocycle of bacteriorhodopsin**

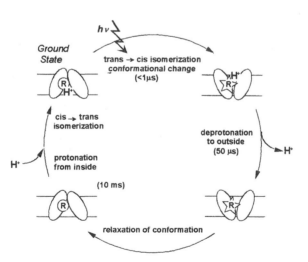

Upon illumination, the bound retinal undergoes a *trans* to *cis* isomerisation which induces a protein conformational change. The new environment of the Shiff base changes its p$K_a$ and a proton leaves to the outside. Relaxation of the protein conformation is accompanied by reprotonation from the interior and a transition back to the *trans* form of the retinal. The conformational changes are slight and are exaggerated in the figure.

retinal. On illumination, the all-*trans* form of retinal is converted to the 13-*cis* form with a shift in geometry of the protonated Schiff base. The result of is to lower the $pK_a$ of the
nitrogen in the retinal ($pK_a$ of around 10) and cause the
release of a proton to the outside medium via aspartate and
arginine residues on one of the helices. Corresponding conformational changes in the protein reorientate the retinal and
restore the high $pK_a$ of the Schiff base. The retinal then repro-
tonates by picking up a proton from the cytosolic side of the
membrane before converting back to the all-*trans* form. The
change in protein conformation plays a crucial part in the
photocycle by acting as a molecular switch to reorient the
deprotonated Schiff base from the extracellular side to the
cytoplasmic side of the membrane.

An observed protonmotive force of around 180 mV is
equivalent to 17.4 kJ for each mole equivalent of charge
transferred across the plasma membrane by the proton
pump (see Equation (7.9)) which provides more than enough
energy for a stoichiometry of one $H^+$ translocated for each
photon absorbed by the retinal pigment (see Table 8.1). The
protonmotive force is used not only to generate ATP but also
to actively extrude sodium ions, which are always leaking
into the cell from the high outside salt concentration. The
internal potassium chloride concentration has to be kept
high to maintain osmotic equilibrium and this involves the
action of a second light-driven pump, *halorhodopsin*, this
time specific for chloride ions.

Isolated aggregates of bacteriorhodopsin from *H. halobium*
were used by Racker and Stoeckenius in a historic experiment in 1974 to test the chemiosmotic theory of Mitchell.
Liposomes were prepared with bacteriorhodopsin and
$F_0F_1$-ATP synthase incorporated across the lipid bilayer.
Illumination under appropriate conditions produced ATP
from ADP and phosphate, thus showing the ability of a proton gradient to couple the favourable free energy change
from the light reaction to the unfavourable phosphorylation
reaction (Figure 8.4). One interesting point about the experiment was the successful use of a combination of material
from three kingdoms, plant (lipid from soybean), animal
($F_0F_1$-ATP synthase from beef heart) and archaebacteria
(bacteriorhodopsin from *H. halobium*).

### 8.2.2 *Bacteriochlorophyll-containing systems*

Chemically, chlorophylls are derived from porphyrin (see
Figure 4.1) and contain $Mg^{2+}$ coordinated to the central nitro-
gens. The pigment systems of photosynthetic bacteria con-

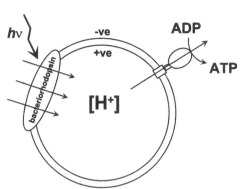

*Figure 8.4*
**Reconstituted proteoliposomes of bacteriorhodopsin and ATP synthase**

Proteoliposomes incorporating bacteriorhodopsin and $F_0F_1$-ATP synthase can synthesise ATP when illuminated. The coupling mechanism is the proton gradient generated by the light-driven proton pump of the bacteriorhodopsin.

tain several bacteriochlorophylls, all differing slightly in the nature of the side-chains attached to the chlorin ring. The substitutions serve to shift the absorption spectra to longer wavelenths in the visible spectrum, even up to 960 nm in the case of bacteriochlorophyll *b* of *Rhodopseudomonas viridis*. Other accessory pigments, for example *bacteriopheophytin* and *carotenoids*, are also found in photosynthetic systems. The advantage is that overlap of different pigments increases the spectral range for light absorption and increases photosynthetic efficiency.

The first stage of energy capture involves *light-harvesting complexes* (LHCs) whose main function is to transfer solar energy to the *photoreaction centres* where the actual redox reactions occur (Figure 8.5). Excitation of electrons in bacteriochlorophyll molecules of the LHC is followed by a rapid return of the electrons to their ground state and the transfer of energy to a neighbouring bacteriochlorophyll. The conjugated double bond system of the chlorin ring provides a system of delocalised energy levels in the visible and far red spectrum. The different side-chains in the various chlorophyll molecules and their interaction with protein in the antenna spread the absorption bands over a range of wavelengths. *Resonance energy transfer* takes place by a non-radiative process (Figure 8.6) and depends strongly on the intermolecular distance and the orientation of the ring systems. It is essential that the pigments are in close contact, otherwise transfer efficiency falls. The collected energy can be transferred over considerable distance to the reaction centre with high speed ($>10^{-12}$ s) and little energy loss. Bacterial

*Figure 8.5*
**The role of light-harvesting complexes**

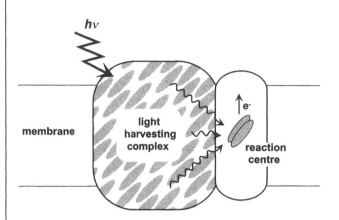

*hv*

membrane

light harvesting complex

e⁻

reaction centre

The light-harvesting complexes are pigment-protein assemblies in the photosynthetic membrane that collect solar energy and deliver it to reaction centres where electron excitation occurs. A variety of membrane antenna complexes exist in bacteria, algae and green plants, reflecting the range of ecological niches of these photosynthetic organisms.

*Figure 8.6*
**Energy transfer by inductive-dipole resonance**

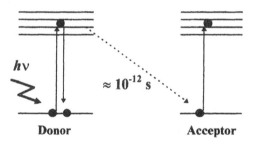

*hv*

$\approx 10^{-12}$ s

**Donor**                    **Acceptor**

Electron excitation in closely packed molecules can lead to dipole–dipole interaction if the dipoles are in the correct orientation for energy transfer by the *Förster mechanism*. Fast energy transfer minimises the vibrational energy loss. The chlorophyll molecules have to be correctly orientated to each other and at close distance (<10 nm) for efficient transfer.

Pheophytins are identical to chlorophylls but lack the central magnesium ion.

LHCs vary in structure and composition to fit different ecological niches. A typical complex may contain 10–500 antenna bacteriochlorophyll molecules complexed with protein.

Most of our structural knowledge of photoreaction centres comes from the crystal structure of the centres from the purple bacteria *Rhodopseudomonas viridis* and *Rhodopseudomonas sphaeroides*. The complex from *R. viridis* was the first to be crystallized and its structure was determined by x-ray crystallography by Deisenhofer, Michel and Huber (who shared the Nobel Prize for chemistry in 1988). It is a transmembrane complex of four polypeptide subunits associated with four haem groups, four bacteriochlorophyll *b* molecules, two *bacteriopheophytins*, two bound quinones and an iron atom. There is twofold symmetry but strangely (at least at the present time) electrons flow only along one of the two branches. The initial photochemical event is the excitation and ejection of an electron from a special pair of bacteriochlorophylls at the reaction centre. The electron is captured by a pheophytin molecule and transfered to one of the two quinone molecules $Q_A$ within a few hundred picoseconds. The primary system is now left in an oxidised state and needs to be reduced before the cycle can continue. This is done via the bound haems (*c*-type cytochromes). Another electron from photoexcitation of the reaction centre can then be passed to $Q_B$ along the same pathway as before. Two protons are taken up and the reduced quinone is released to the bulk quinone pool. Electron flow in bacterial systems is cyclic (Figure 8.7) and the electrons return to the reaction centres via a cytochrome $bc_1$ complex (see Section 4.3). Free energy is transduced into an electrochemical gradient of protons by means of a Q-cycle (see Section 7.5).

Electron flow in bacterial systems does not directly produce reducing equivalents for fixation of $CO_2$. As well as driving ATP synthesis, the protonmotive force can be used

**Figure 8.7**
**Bacterial photosynthesis**

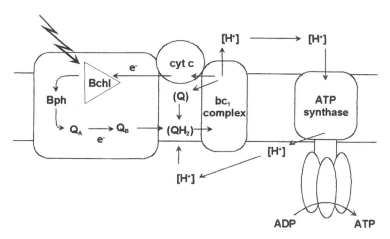

The process is cyclic and involves energy capture by pigment molecules and transfer to a photoreaction centre. Excitation of bacteriochlorophyll in the centre causes electron flow to a quinol pool in the membrane. Electrons are returned to the reaction centre via a cytochrome $bc_1$ complex and *c*-type cytochromes. Free energy is conserved in the form of proton gradient by a Q-cycle mechanism. This is then used to drive $NAD_{red}$ formation as well as ATP synthesis. Bchl, bacteriochlorophyll; Bph, bacteriopheophytin; Q, quinone.

to reverse electron transport in complex I ($NAD_{red}$/quinone reductase) to produce $NAD_{red}$. $NADP_{red}$ can be formed from $NAD_{red}$ by the action of a *transhydrogenase*, also powered by the protonmotive force.

$$NADP_{ox} + NAD_{red} \xrightarrow{\Delta\mu_{H^+}} NADP_{red} + NAD_{ox} \qquad (8.2)$$

The source of electrons for NAD and subsequently carbon dioxide reduction varies between different bacteria. The purple sulphur and green sulphur bacteria utilize reduced sulphur compounds such as hydrogen sulphide. Overall the process can be represented as

$$CO_2 + 2H_2S \rightarrow \{HCOH\} + 2S^0 + H_2O \qquad (8.3)$$

The sulphides are already strong reductants (see Table 4.2) and the free energy needed for NAD and NADP reduction is relatively small. These anaerobes can therefore exist in environments where light intensities are low and the light is in the long-wavelength region of the spectrum. These conditions abound in the deep anoxic layers of stagnant waters. The purple non-sulphur bacteria, Rhodospirillaceae, can oxidise organic compounds for energy with oxygen as the terminal acceptor in the absence of light. When food is scarce they are able to operate photosynthetically under anaerobic conditions, oxidising the end products of fermentation such as molecular hydrogen from other bacteria as a source of electrons.

### 8.2.3 *Chlorophyll-containing systems*

Green plants, algae and cyanobacteria (blue-green algae) use water as a source of electrons, releasing molecular oxygen as a waste product. The overall process can be compared with bacteriochlorophyll systems (Equation (8.3)).

$$CO_2 + 2H_2O \rightarrow \{HCOH\} + O_2 + H_2O \qquad (8.4)$$

It was the gradual evolution of water-splitting photosynthetic systems that lead to extensive changes in the earth's atmosphere around two billion years ago (see Box 8.2).

The photosynthetic apparatus of green plants and algae is arranged in *chloroplast* organelles (Figure 8.8). Internally, the chloroplast contains stacks of *thylakoid membranes* called *grana* in which are embedded the light-sensitive pigments and the photoreaction centres. The membranes also contain the various electron transport components which are involved in the transfer of electrons from water to NADP. The internal soluble phase of the chloroplast is known as

*Box 8.2* **Evolution of the atmosphere**

The atmosphere of the early earth was anaerobic. The major components were carbon dioxide, nitrogen, methane and carbon monoxide, probably in the approximate ratios of 10:1:<0.1:<0.1 with smaller concentrations of hydrogen, ammonia and hydrogen sulphide. The mean redox potential would be below 0.0 V. Atmospheric oxygen only arose with the development of water-splitting photosynthetic organisms on a large scale. We know roughly when this happened from studying the geology of banded iron formations. Ferrous iron ($Fe^{2+}$) is quite soluble (around 0.1 M at pH 7). This contrasts with the oxidised form of iron ($Fe^{3+}$), which is essentially insoluble ($10^{-18}$ M at pH 7). The ferrous iron in the early oceans would have precipitated when the dissolved oxygen increased and we can see the effects as red bands of ferric minerals in rocks aged betwen 2.3 to 1.8 billion years ago. Eventually, the natural geochemical buffers for oxygen became saturated and around 1.8 billion years ago aerobic conditions were established in the atmosphere. The subsequent formation of ozone, which absorbs ultraviolet light, allowed terrestrial life to develop (Table 8.3).

**Table 8.3** Evolution of the biosphere

| Age (billions of years) | Events |
| --- | --- |
| 4.5 | formation of earth |
| 3.5 –3.0 | development of anaerobic bacteria |
| 3.0 | photosynthetic bacteria (anaerobic) |
| 2.5–2.0 | aerobic photosynthesis and aerobic respiration |
| 2.2–1.8 | growth of oxygen-rich atmosphere |
| 1.5–1.8 | origin and development of eukaryotes |

the stroma and contains all the necessary enzymes for the fixation of carbon dioxide into organic carbon.

Light-harvesting complexes of chlorophyll-containing systems are large and abundant in green plants. The most common, named LHC-II, is the main collector of solar energy in the biosphere. The monomer comprises three membrane-spanning helices associated with a minimum of 12 chlorophylls and various associated pigments such as *carotenoids* plus two different lipids, *phosphatidyl glycerol* and *digalactosyl diacylglycerol*. The LHC-II is associated into trimers *in situ*. Rapid collection and transfer of photon energy to the photoreaction centres occurs by resonance energy transfer as in the bacteriochlorophyll systems. In green plants, algae and cyanobacteria, however, two distinct types of photoreaction centre act in cooperation (Figure 8.9). A photosystem which can operate up to wavelengths of light of 700 nm has been termed *photosystem I* (PS-I). A second system, with structural similarities to the photosystems of purple bacteria, operates up to slightly lower wavelengths (680 nm) and is named *photosystem II* (PS-II). Energy focused onto PS-II by

## Figure 8.8
**The chloroplast**

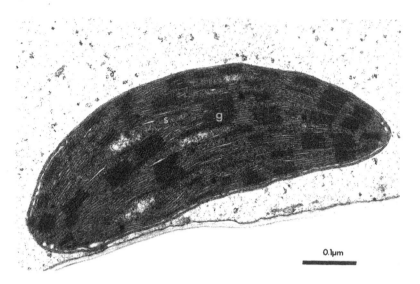

Electron micrograph of a plant cell showing chloroplast with grana stacks (g). (Courtesy John Pacy, Electron Microscope Unit, King's College London.)

## Figure 8.9
**Photosynthesis by green plants, algae and cyanobacteria**

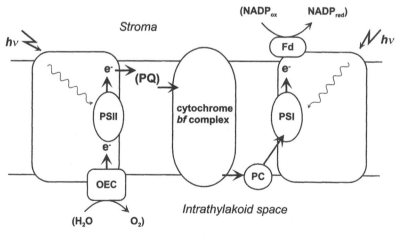

The two types of photosystems are embedded in the photosynthetic membranes as separate complexes. Electron transfer between the photosystems involves plastoquinone (PQ), a cytochrome *bf* complex, and the water-soluble copper-containing protein plastocyanin (PC). A protonmotive force is generated by a Q-cycle mechanism. The electrons are finally used to reduce NADP via ferredoxin (Fd). Re-reduction of photosystem II is mediated by a water-splitting complex (or oxygen-evolving complex, OEC) containing manganese.

the LHC-II results in electron excitation (leaving the centre around +1.1 V) and electron transfer to plastoquinone via pheophytin. The cytochrome *bf* complex is analogous to the bacterial and mitochondrial $bc_1$ complex and produces a protonmotive force by a Q-cycle mechanism (see Section 7.5). Electrons are passed to oxidised PS-I (at around + 0.5 V) by a water-soluble copper-containing protein, *plastocyanin* ($E^{\circ\prime} = +0.34$ V). The electron ejected from PS-I by photon energy is used to reduce NADP ($E^{\circ\prime} = -0.32$ V) after transport through another water-soluble protein, *ferredoxin* ($E^{\circ\prime} = -0.43$ V), whose redox centre contains a 2Fe-2S iron–sulphur complex (see Figure 4.1). Overall, the redox difference between the oxidised PS-II and the primary electron acceptor of PS-I is around 2 V. From Table 8.1 it can be seen that even with 100% efficiency it would need more than one photon of red light to excite an electron over this potential. The two photosystems acting in concert enable the full redox range to be covered.

The redox potential of oxidised PS-II is sufficient to drive the oxidation of water ($E^{\circ\prime} = +0.82$). To carry out this reaction, the *water-splitting complex* has to abstract four electrons from two water molecules to produce one oxygen molecule. During the reaction, the release of damaging oxygen intermediates is avoided by the use of a cluster of four manganese atoms to accumulate four positive charges before dioxygen is released. A tyrosine radical acts as an intermediary to transfer electrons to the oxidised PS-II.

The electron from PSI can also cycle back to re-reduce the oxidised chlorophyll molecule in PSI using the same electron transport chain as in non-cyclic flow. No $NADP_{red}$ is made but ATP is still produced. The process is known as *cyclic phosphorylation*. We shall see in the next section that the additional ATP is required for carbon dioxide fixation.

## 8.3 Carbon dioxide fixation

### 8.3.1 *Ribulose-1,5-bisphosphate carboxylase*

The chloroplast stroma contains the necessary enzymes to catalyse carbon dioxide fixation in the dark, once a supply of ATP and $NADP_{red}$ is present. The so-called *dark reactions* start with the initial capture of carbon dioxide by *ribulose-1,5-bisphosphate carboxylase* (often termed *Rubisco*), probably the most abundant protein in the biosphere. The enzyme, found also in photosyntheic bacteria, catalyses the carboxylation and subsequent hydrolysis of ribulose 1,5-

The primary product of the dark reactions of photosynthesis in $C_3$ plants is 3-phosphoglycerate.

bisphosphate, a favourable reaction under standard conditions ($\Delta G^{\circ\prime} = -51.9\,\mathrm{kJ\,mol^{-1}}$).

$$
\begin{array}{c}
CH_2OPi \\
| \\
C=O \\
| \\
H-C-OH \\
| \\
H-C-OH \\
| \\
CH_2OPi
\end{array}
\quad
\xrightarrow{\quad CO_2 \quad}
\quad
2\times
\quad
\begin{array}{c}
CH_2OPi \\
| \\
H-C-OH \\
| \\
COO^-
\end{array}
$$

**ribulose 1,5-bisphosphate**          **3-phosphoglycerate**

(8.5)

Unfortunately, Rubisco has a relatively low affinity for carbon dioxide ($K_m \approx 10\text{--}20\,\mu M$) when compared with the normal carbon dioxide concentrations in the cell (around 10 μM). In the presence of normal oxygen concentrations (around 250 μM), oxygen competes with carbon dioxide and Rubisco will catalyse the reaction of ribulose 1,5-bisphosphate with molecular oxygen at the same catalytic site as the carboxylation. One of the products of the reaction is 2-*phosphoglycolate*, a two-carbon compound. This is effectively a waste product as far as metabolism is concerned. Some of the carbon can be salvaged by conversion into *glycolate* which leaves the chloroplast and is oxidised to *glyoxylate* in peroxisomes. Transamination then gives glycine (a two-carbon amino acid), two molecules of which can then form serine (a three-carbon amino acid) with the loss of one carbon as carbon dioxide. The overall pathway consumes oxygen and releases carbon dioxide with a waste of free energy, a process known as *photorespiration*. At concentrations of carbon dioxide in the cell, up to a quarter of the fixed carbon can be lost by photorespiration. The structure of Rubisco has now been determined by x-ray crystallography and attempts are being made to genetically design an enzyme with an increased carboxylation efficiency. Rubisco is activated by carbamylation (reaction of a side-chain lysine group with carbon dioxide) and the addition of a magnesium ion.

The relatively low affinity of Rubisco for carbon dioxide and the competing reaction with oxygen point to an early evolution for this enzyme. It is likely that primitive photosynthetic organisms used Rubisco before atmospheric oxygen grew to significant levels (see Box 8.2) when the earth's atmosphere contained more carbon dioxide.

$$
\begin{array}{c}
COO^- \\
| \\
CH_2OH
\end{array}
$$
**glycolate**

$$
\begin{array}{c}
\searrow O_2 \\
\Big\downarrow \\
\searrow H_2O_2
\end{array}
$$

$$
\begin{array}{c}
COO^- \\
| \\
C \\
O^{\diagup\,\diagdown}H
\end{array}
$$
**glyoxylate**

### 8.3.2 *The Calvin cycle*

Carbon dioxide assimilation by the *Calvin cycle* proceeds by a reversal of part of the glycolytic pathway (see Section 5.4) with the formation of glyceraldehyde 3-phosphate from 3-phosphoglycerate (the reductive phase of assimilation). This part of the sequence requires some of the ATP and $NADP_{red}$ made by the light reactions of photosynthesis.

$$3\text{-phosphoglycerate} + ATP \rightarrow$$
$$1,3\text{-bisphosphoglycerate} + ADP \tag{8.6}$$

$$(\Delta G^{\circ\prime} = +18.8\,\text{kJ}\,\text{mol}^{-1})$$

$$1,3\text{-bisphosphoglycerate} + NADP_{red} \rightarrow$$
$$\text{glyceraldehyde 3-phosphate} + NADP_{ox} \tag{8.7}$$

$$(\Delta G^{\circ\prime} = -6.3\,\text{kJ}\,\text{mol}^{-1})$$

Equations (8.6) and (8.7) express the main energetic stages of the dark reactions. The carbon oxidation state has been changed from the +4 state in carbon dioxide to +1 in an aldehyde (see Table 5.1). The redox potential of the glyceraldehyde/3-phosphoglycerate couple is $-0.55$ V (see Table 4.2).

In theory, the triose phosphates could then be used to synthesise glucose as in gluconeogenesis. However, a straightforward linear pathway would not regenerate any ribulose 1,5-bisphosphate and carbon dioxide fixation would soon stop. A cyclic pathway which synthesises a six-carbon sugar and regenerates phosphorylated ribulose was worked out in the late 1940s by Calvin (Nobel Prize for chemistry in 1961) and others, notably Benson and Bassham. Advantage was taken of the newly available carbon isotopes to trace the formation of intermediates following exposure of photosynthesising algae to $^{14}CO_2$. After the triose phosphate is formed, ribulose 1,5-bisphosphate is regenerated by a series of interconversion reaction involving various sugar phosphates (Figure 8.10). Six turns of the cycle fix six carbon dioxide molecules into one six-carbon sugar and regenerate six phosphorylated ribulose molecules. One extra ATP is needed in the final stage in order to phosphorylate ribulose 5-phosphate. The overall stoichiometry of $NADP_{red}$ to ATP is 3:2 and the extra ATP is supplied by cyclic phosphorylation (see previous section).

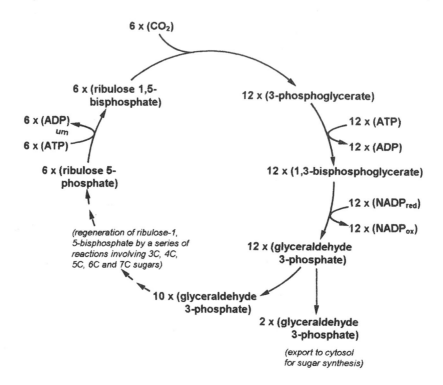

*Figure 8.10*
**The photosynthetic carbon reduction (Calvin) cycle**

The initial fixation of carbon dioxide in $C_3$ plants generates two molecules of 3-phosphoglycerate from one molecule of carbon dioxide and one molecule of ribulose 1,5-bisphosphate. The 3-phosphoglycerate is reduced to glyceraldehyde 3-phosphate at the expense of ATP and $NADP_{red}$ generated in the light reactions. The cycle has to function six times to fix six molecules of carbon dioxide into two triose phosphates for export into the cytoplasm. The remaining ten triose phosphates are used to regenerate six molecules of ribulose 1,5-bisphosphate to keep the cycle turning.

$6CO_2$ + 6 ribulose 1,5-bisphosphate + $12NAD_{red}$ + 18ATP
  (1C)                                    (5C)

$$\downarrow \qquad (8.8)$$

glucose + 6 ribulose 1,5-bisphosphate + $12NAD_{ox}$ + 18ADP
  (6C)                        (5C)
                                                    + 18 phosphate

Glucose can be further metabolised to starch in the chloroplast. Alternatively, triose phosphates can be exported into the cytosol for sucrose synthesis. Remember that five out of every six triose phosphates would have to remain in the chloroplast to regenerate ribulose 1,5-bisphosphate. The exported triose phosphates are converted to glucose 6-phosphate by reverse glycolytic reactions. Sucrose, a very soluble disaccharide, is one of the main storage sugars of plants. However there is an energy cost of storage. Linking glucose and fructose together is an unfavourable reaction under standard conditions:

$$\text{glucose} + \text{fructose} \rightarrow \text{sucrose} + H_2O \quad (\Delta G^{\circ\prime} = 30\,\text{kJ}\,\text{mol}^{-1})$$

(8.9)

The reaction therefore has to be coupled to a more favourable one. In this case the hydrolysis of UTP ($\Delta G^{\circ\prime} = -31\,\text{kJ}\,\text{mol}^{-1}$). The reaction sequence is

$$\text{UTP} + \begin{array}{c}\text{glucose}\\ \text{1-phosphate}\end{array} \longrightarrow \text{pyrophosphate} + \boxed{\text{UDP-glucose}} \quad (8.10)$$

$$\boxed{\text{UDP-glucose}} + \begin{array}{c}\text{fructose}\\ \text{6-phosphate}\end{array} \longrightarrow \text{UDP} + \begin{array}{c}\text{sucrose}\\ \text{6-phosphate}\end{array} \quad (8.11)$$

UDP-glucose acts as the common intermediate linking the two reactions. The phosphate on sucrose 6-phosphate is removed by simple hydrolysis.

### 8.3.3 The Hatch–Slack (C₄) pathway of carbon dioxide fixation

In many tropical plants and grasses, the initial product of carbon dioxide fixation is not the $C_3$ metabolite, 3-phosphoglycerate, but a $C_4$ metabolite, oxaloacetate. Plants with this route for carbon dioxide fixation are known as *C₄ plants*. The light reactions take place in the same way as in $C_3$ plants with the production of $NADP_{red}$ and ATP. However, the leaves of $C_4$ plants have additional specific enzymes which are located in two cell types. The *mesophyll cells* are located close to the stomatal entrance so that the bicarbonate in their cytosol is in equilibrium with atmospheric carbon dioxide. *Carbonic anhydrase* catalyses a rapid conversion of carbon dioxide to bicarbonate:

---

The primary product of the dark reactions of photosynthesis in C₄ plants is oxaloacetate.

---

$$CO_2 \text{ (dissolved gas)} + H_2O \rightarrow HCO_3^- + H^+ \quad (8.12)$$

The equilibrium constant for this reaction is $4.4 \times 10^{-7}$ ($pK_a = 6.1$). From the *Henderson–Hasselbalch equation* the following relationship can be written:

$$pH = pK_a + \log \frac{[HCO_3^-]}{[CO_2]} \quad (8.13)$$

Therefore, at a cytosolic pH of 7.4, the ratio of bicarbonate to dissolved carbon dioxide will be around 20:1. *Phosphoenolpyruvate carboxylase* catalyses the formation of oxaloacetate from bicarbonate and phosphoenol pyruvate, a very favourable reaction under standard conditions:

$$phosphoenolpyruvate + HCO_3^- \rightarrow oxaloacetate + phosphate$$

(8.14)

$$(\Delta G^{\circ\prime} = -30\,kJ\,mol^{-1})$$

This method of fixing carbon dioxide has advantages over the Rubisco reaction since the phosphoenol-pyruvate carboxylase reacts with bicarbonate which is in equilibrium with carbon dioxide at very low concentrations (<1 µM). The enzyme is also insensitive to molecular oxygen, unlike Rubisco which 'wastes' a significant fraction of its activity in photorespiration.

The oxaloacetate is reduced to malate using the $NADP_{red}$ from the normal light reactions. The malate is transported to the bundle-sheath cells, where it is decarboxylated to pyruvate, releasing carbon dioxide (Figure 8.11). Thus atmospheric carbon dioxide has been delivered by a carbon dioxide 'pump' from relatively low concentrations in the

---

*Figure 8.11*
**Carbon fixation by C$_4$ plants**

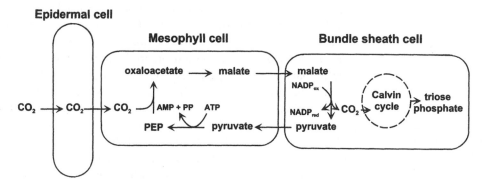

Carbon dioxide is first fixed as oxaloacetate in the mesophyll cells. It is then passed to the bundle-sheath cells as malate (in some species aspartate is the diffusing metabolite). Once there, malate is decarboxylated to pyruvate, releasing the carbon dioxide at a high local concentration for capture by Rubisco and metabolism by the Calvin cycle. The pyruvate diffuses back to the mesophyll cells for conversion back to phosphoenol pyruvate (PEP). The advantage to C$_4$ plants of 'pumping' carbon dioxide to specialised cells is that the higher concentrations of carbon dioxide increase the carbon fixing efficiency of Rubisco. In C$_3$ plants much energy is lost by Rubisco oxygenase activity (photorespiration).

mesophyll cells to a higher concentration in the bundle-sheath cells. The liberated carbon dioxide can then be assimilated efficiently into organic carbon by Rubisco and the Calvin cycle. The higher carbon dioxide concentrations also help to decrease the wasteful oxygenase activity of Rubisco. To complete the carbon dioxide 'pump-cycle', pyruvate has to return to the mesophyll cell to be converted back to phosphoenol pyruvate. The conversion is an unfavourable reaction and requires coupling to ATP hydrolysis. Nevertheless, this is a price $C_4$ plants are willing to pay for significantly higher growth rates compared with $C_3$ plants.

There is some variation between $C_4$ plants in the nature of the compound transported between the mesophyll and bundle-sheath cells. Some use aspartate rather than malate, but

*Box 8.3* **Biomass energy**

Biomass is a measure of the amount of organic carbon in the biosphere, excluding carbon deposits of coal and oil. It supplies around 14% of the global energy for human use (Figure 8.12). Coal, oil and natural gas provide the bulk of the remainder. The advantage of energy from biomass is that it has the potential of providing a renewable source with little or no net carbon dioxide emission to the atmosphere (Figure 8.13). The carbon dioxide from the burning of plant material can be recycled back into biomass over a relatively short time by photosynthesis. In contrast, the use of fossil fuels, laid down over a period of several million years, results in a net contribution to atmospheric carbon dioxide. Biomass also has the favourable property of a low sulphur content compared to most fossil fuels.

Cellulose, a representative product of photosynthesis, has a carbon content of 44% and a fuel value of around $16 \, MJ \, kg^{-1}$ dry weight (see Table 2.1). Oils from plant seeds are much higher in energy content, around 40 $MJ \, kg^{-1}$. At present, the main forms of biomass energy are wood and charcoal for fuel. Increasingly however, biomass is being converted to other more convenient fuels such as ethanol and methane. Ethanol is blended with gasoline in several countries, especially in Brazil where sugar cane is used as a primary source of biomass. Various carbohydrates such as corn and sugar are readily converted to ethanol by fermentation in the presence of yeast. Cellulose-based material is more difficult to use and first requires partial hydrolysis

before efficient fermentation can take place. Methane can be made from most plant and animal waste by the action of mixed populations of anaerobes (see Section 5.6). Yields can reach up to $0.3 \, m^3$ of gas per kg of added biomass with thermal efficiency conversions of around 40% of the initial substrate being maintained in the final methane.

A few simple calculations can give some idea of the annual global production of biomass energy. Remembering that 1 metric tonne is equivalent to 1000 kg, the heating value of dry carbohydrate is equivalent to $16 \times 10^9 \, J \, tonne^{-1}$. The amount of organic carbon fixed by terrestial photosynthesis every year is around $120 \times 10^9$ tonnes, which is equivalent to $19.2 \times 10^{20}$ J per year (from a total solar energy flux on to the surface of the earth, in the wavelength range used by photosynthesis, of $3 \times 10^{24}$ J per year). This is more than six times the present use of fossil fuels. Of course only a fraction of biomass growth could be harvested, but even using 8% of the annual photosynthetic production could halve the present use of fossil fuels. Unfortunately, the environmental advantage of using biomass energy is lost if an equal amount of new biomass is not created by photosynthesis. Quick-growing, high-yield fuelwood and other energy crops such as sugar cane are available for fast recycling, but a major destruction of biomass occurs every year by deforestation. An estimated 200 million tonnes of net biomass is lost each year by destruction and non-replacement of forests areas.

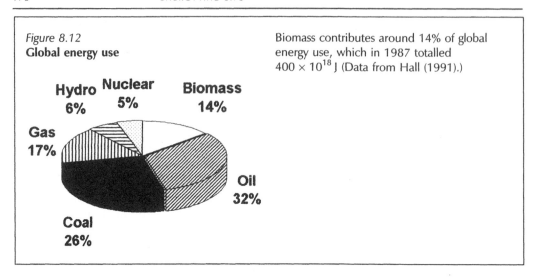

*Figure 8.12*
**Global energy use**

Biomass contributes around 14% of global energy use, which in 1987 totalled $400 \times 10^{18}$ J (Data from Hall (1991).)

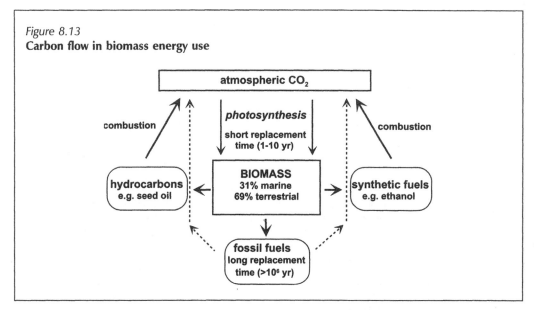

*Figure 8.13*
**Carbon flow in biomass energy use**

the net effect is still to deliver carbon dioxide at high local concentrations for Rubisco activity. Another major modification found in many succulent plants, including the family Crassulaceae, is termed *crassulacean acid metabolism* (CAM). Carbon dioxide is initially captured by the $C_4$ pathway described above, but at night using the $NADP_{red}$ and ATP accumulated during the previous daytime. The $C_4$ compounds are stored until the next day when they are then decarboxylated in the same cells to release carbon dioxide for Rubisco activity. Why should the CAM species separate the carbon dioxide-trapping pathway in time rather than in space? The reason is that these plants usually inhabit very arid regions and maintain closed stomata during the day to minimise water loss. It is only at night when the stomata open that carbon dioxide can freely diffuse to the underlying photosynthetic tissue.

### 8.3.4 Carbon dioxide growth-dependence of $C_3$ and $C_4$ plants

Since carbon dioxide is delivered to Rubisco at high local concentrations by the $C_4$ pathway, the $C_4$ plants are relatively insensitive to increases in atmospheric carbon dioxide. Not so the $C_3$ plants. For many years it has been known that burning propane in greenhouses (for heating) also had the additional effect of releasing carbon dioxide. Plants such as tomatoes increased in yield. Comparing the rates of photosynthesis between $C_3$ and $C_4$ plants shows that $C_3$ plants respond to carbon dioxide concentrations far more than $C_4$ plants, who have their own built-in carbon dioxide concentrating system. It must be remembered, however, that carbon dioxide is only one of the factors that may limit growth. Moderate temperatures, nutrient supply and inadequate light also constrain growth efficiency. Experiments in seeding the surface waters of equatorial oceans with low concentrations of dissolved iron has been shown to trigger massive phytoplankton blooms which consume large quantities of carbon dioxide and nitrate. In this case growth appears to be limited by iron bioavailability. It has been suggested that increased iron availability to the southern oceans during glacial times could be responsible for the observed low atmospheric levels of carbon dioxide at that time.

## Summary

- Over 99% of the free energy of the biosphere arises from the sun. Radiant energy interacts with matter in the form of photons with discrete energy values. For interaction to occur, the energy of the photon has to exactly match an energy gap between two states of the absorbing molecule.

- There are three main systems for photon capture based on the type and organisation of the light-absorbing pigments. Bacteriorhodopsin-containing systems directly transduce the light-energy into the energy of an electrochemical gradient of protons using a light-driven proton pump. The free energy of the gradient is then coupled to other biosynthetic reactions.

Bacteriochlorophyll-containing systems are found in the green and purple bacteria, which are photosynthetic anaerobes using reductants such as hydrogen sulphide or partially reduced organic compounds such as malate as a source of reducing equivalents. Chlorophyll-containing systems are found in green plants, algae and cyanobacteria. These use water as a source of electrons to reduce carbon dioxide to organic material. Oxygen is released as a waste product.

- Carbon dioxide is initially fixed into organic carbon either as the three-carbon compound 3-phosphoglycerate, in $C_3$ plants, or as the four-carbon compound oxaloacetate, in $C_4$ plants. $C_3$ plants convert the 3-phosphoglycerate

to triose phosphates and then to hexose sugars by the action of the Calvin cycle. $C_4$ plants convert the oxaloacetate to malate to be transported from the mesophyll cells, located close to the stomatal entrance to the bundle-sheath cells where decarboxylation of pyruvate releases carbon dioxide. By this mechanism, atmospheric carbon dioxide is delivered from relatively low concentrations in the mesophyll cells to a higher local concentration in the bundle-sheath cells. The liberated gas can then be trapped efficiently by the enzyme ribulose-1,5-bisphosphate carboxylase (Rubisco) and assimilated using the Calvin cycle. The higher carbon dioxide concentration helps to decrease the wasteful oxygenase activity of Rubisco.

## Selected reading

Castresana, J. and Saraste, M., 1995, Evolution of energetic metabolism: the respiration-early hypothesis, *Trends Biochem. Sci.* **20**, 443–448. (An analysis questioning conventional ideas on the origin of aerobic respiration)

Coale, K.H. *et al.*, 1996, A massive phytoplankton bloom induced by an ecosystem-scale iron fertilization experiment in the equatorial pacific ocean, *Nature* **383**, 495–501. (An interesting large-scale experiment concerned with the role of oceans as a source and sink for atmospheric carbon dioxide)

Hall, D.O., 1991, Biomass energy, *Energy Policy* **19**, 711–737. (A useful survey of the economic, social and environmental issues of biomass energy use)

Hall, D.O. and Rao, K.K., 1994, *Photosynthesis*, Cambridge: Cambridge University Press. (A clear and authoritative account of the subject)

Lanyi, J.K., 1994, Bacteriorhodopsin as a model for proton pumps, *Nature* **375**, 461–463. (A comparison of proton pumping mechanisms between bacteriorhodopsin, cytochrome oxidase and $F_0F_1$-ATP synthase.)

Racker, E. and Stoeckenius, W., 1974, Reconstitution of membrane vesicles catalysing light-driven proton uptake and adenosine triphosphate formation, *J. Biol. Chem..* **249**, 662–663. (A classic experiment that helped to establish the chemiosmotic theory)

### Structures

Deisenhofer, J. and Michel, H., 1991, High-resolution structures of photosynthetic reaction centres, *Annu. Rev. Biophys. Biophys. Chem.* **20**, 247–266.

Henderson, R., Baldwin, J.M., Ceska, T.A., Zemlin, F., Beckmann, E. and Downing, K.H., 1990, Model for the structure of bacteriorhodopsin based on high resolution electron cryo-microscopy, *J. Mol. Biol.* **213**, 899–929.

Kühlbrandt, W, Wang, D.G., Fujiyoshi, Y., 1994, Atomic model of plant light-harvesting complex by electron crystallography, *Nature* **367**, 614–621. (A 3.3 Å map of LHC-II derived from electron cryo-microscopy of two dimensional crystals)

Schneider, G., Linqvist, Y. and Brändén, C.-I., 1992, Rubisco: structure and mechanism, *Annu. Rev. Biophys. Biomol. Struct.* **21**, 119–143.

## Study problems

1.  How many watts would be equivalent to a flux of 400 nm photons falling onto the suface of the earth at a rate of $1.3 \times 10^{21}$ photons $s^{-1}$? What would this value be for light of 700 nm?

2.  Calculate the theoretical minimum number of quanta of light of wavelength 680 nm required to fix 1 mole of carbon dioxide to carbohydrate under standard conditions, given that the standard free energy of oxidation of glucose to carbon dioxide and water is 2870 kJ mol$^{-1}$ (refer to Table 8.1)? Why is this number much greater in practice?

3.  The amount of solar energy in the photosynthetic range striking the earth's surface every day is $4 \times 10^{21}$ J. Calculate the mass in tonnes of carbon dioxide fixed by photosynthetic organisms assuming a capture efficiency of 0.3%. (Refer to Section 8.1.)

# Answers to Numerical Problems

## Chapter 2

1. 156 kJ
2. 0.72
3. 1962 J
4. 318 m
5. (a) 3 min; (b) 37 min
6. (a) 8316 kJ; (b) 5.2 kg; (c) 900 kJ

## Chapter 3

1. (a) 9.9; (b) $-5.7\,\mathrm{kJ\,mol^{-1}}$
2. $-15.3\,\mathrm{kJ\,mol^{-1}}$
3. $5.5\ \mathrm{mmol\ s^{-1}}$
4. Refer to Equations (3.3) and Sections (3.3) and (3.4).
5. (a) Yes, since the overall $\Delta G^{\circ\prime}$ is $-10.5\,\mathrm{kJ\,mol^{-1}}$. (b) No, since the overall $\Delta G^{\circ\prime}$ is $+56.3\,\mathrm{kJ\,mol^{-1}}$
6. $32.9\,\mathrm{kJ\,mol^{-1}}$

## Chapter 4

1. (a) Biochemistry texts usually use $E^{\circ\prime}$ rather than $E^{\circ}$. The activity of $H^{+}$ at pH 7 is defined as 1 rather than $10^{-7}$. (b) $E = -0.059\,\mathrm{pH}$
2. $-0.29$ V
3. $84.9\,\mathrm{kJ\,mol^{-1}}$
4. (a) $+0.31$ V; (b) $+0.37$ V; (c) 90% inhibition
5. 0.01% reduced
6. 0.156 J

## Chapter 5

1. The carbonyl carbon changes from $+3$ to $+4$ and the $\alpha$-carbon changes from $-2$ to $-3$.

2.  807

3.  626

4.  The reactions catalysed by citrate synthase and fumarase involve molecular water. The conversion of citrate to isocitrate involves successive dehydration and hydration reactions. The reaction catalysed by succinyl-CoA synthase (Equation (5.11)) involves the scavenging of a water equivalent from inorganic phosphate)

5.  10.4 mole (1 from substrate level phosphorylation, 7.8 from the oxidation of $NAD_{red}$ and 1.6 from the oxidation of succinate)

6.  They lack the glyoxylate pathway. The two-carbon acetyl fragment on acetyl-CoA, produced by $\beta$-oxidation, enters the citric acid cycle by the formation of citrate. Two carbons are lost as carbon dioxide during one turn of the cycle before the starting material for gluconeogenesis, oxaloacetate, can be formed.

# Chapter 7

1.  $\sim$0.1 mm

2.  24 atm

3.  (a) 70.6 kJ; (b) 47.5 kJ

4.  109 mV

5.  (a) 5. (b) 3.33. (c) One mole of ATP is formed, per mole of oxygen atoms consumed, by substrate-level phosphorylation during the oxidation of 2-oxoglutarate to succinate. The remaining 2.33 moles of ATP are formed by oxidative phosphorylation during electron transfer from $NAD_{red}$ to oxygen.

6.  Muscle contraction uses ATP. The lowered ATP/ADP ratio stimulates mitochondrial respiration and oxygen consumption in order to make more ATP. Glycolysis is also stimulated and excess pyruvate, which cannot be oxidised by aerobic metabolism, is converted to lactate which diffuses into the bloodstream.

# Chapter 8

1.  (a) 645 W; (b) 369 W

2.  (a) 2.9. (b) Energy loss occurs in energy capture and also by photorespiration.

3.  $1.1 \times 10^{12}$ Tonnes

# Index

Note: Page numbers in **bold** indicate a major discussion of the entry where there is more than one reference. *t* following a page number indicates tabular material; *f* following a page number indicates a figure.

T - #0235 - 071024 - C0 - 254/178/11 - PB - 9780748404339 - Gloss Lamination